Why
Do Pirates Love
Parrots?

Why
Do Pirates Love
Parrots?

An Imponderables® Book

DAVID FELDMAN

Illustrated by Kassie Schwan

Collins

An Imprint of HarperCollins Publishers

HarperCollins books may be purchased for educational, business, or sales promotional use. For information please write: Special Markets Department, HarperCollins Publishers, 10 East 53rd Street, New York, NY 10022.

First Collins paperback edition published 2007

Designed by William Ruoto

Library of Congress data available upon request.

ISBN: 978-0-06-088843-5 (pbk.)
ISBN-10: 0-06-088843-1 (pbk.)

07 08 09 10 11 ID/RRD 10 9 8 7 6 5 4 3 2 1

This book is dedicated to *Imponderables* readers, especially these five, who represent the contributions of tens of thousands:

Susan Sherman Smith, who died tragically in a fire in 2004, was the first person to contribute an Imponderable used in a book, and it's my all-time favorite: "Why Do Women Open Their Mouths When Applying Mascara?"

Joanna Parker started writing warm, witty, and supportive letters from day one. Her mail is still cherished, even if it no longer comes encased in pink envelopes.

Douglas Watkins, Jr. has sent in by far the most Imponderables of any reader, every one of them handwritten. As usual, he earned a free copy of this book.

Ken Giesbers is the decathlete of Imponderables, often finding errors in the Letters section, but also contributing Imponderables, commenting on Frustables, and acting as a source about airplanes.

Bill Gerk, king of the Frustables, died at the age of seventy-five in 2004. After each *Imponderables* book was published, Bill took it upon himself to research every Frustable. He would send several envelopes full of at least thirty to fifty pages of musings, ranging from hard research to imaginative stabs in the dark. His humor and never-ending curiosity were an inspiration, and in the true spirit of *Imponderables*.

Contents

Preface

This book attempts to answer little mysteries of everyday life, tackling subjects ranging from bottle necks to vultures, medicine balls to barbecues, and yes, parrots and pirates. Our goal is simple. To make all Imponderables walk the plank so that we may live in a world free from the nagging conundrums that plague our everyday life. The trouble is that as soon as one Imponderable is vanquished, more mysteries smack us in the face.

We're here for you. Almost all of the Imponderables in our eleven books come from readers, and we offer a free, autographed copy if you are the first person to pose an Imponderable we answer in a book.

You might have noticed that this book is dedicated to our readers. Your ideas are the lifeblood of our enterprise. We identified five readers by name, but we could have just as well cited Dallas Brozik, Debra Allen, Craig Kirkland, Fred Beeman, Ariel Godwin, Gail Dunson, the Itzcowitz family, or scores of other readers whose enthusiasm brightens up our inbox. Thank you readers—this one's for you.

You may notice a few special features in *Why Do Pirates Love Parrots?* Just as some Web sites include a frequently asked questions section, we've devoted a section to Unimponderables, frequently asked irritating questions that may not fit the precise definition of Imponderables, but are posed to us frequently—incessantly, actually. And we've included an Updates section to spread the latest research about some Imponderables discussed in previous books.

You can't enjoy the full *Imponderables* experience without visiting us on the Web at http://www.imponderables.com. In some cases, we've expanded our discussions in this book on the site, have photos to illustrate the text here, and have comments from readers letting us have it. If we have updates on any of the Imponderables in this book, we'll post them online. The *Imponderables* Web site offers a master index of all the *Imponderables* books and a blog written by Dave Feldman—and absolutely no popup or banner ads.

In some discussions here, the book, we've added URLs of other Web sites, places to look for more information or illustrations. Web sites have a nasty habit of changing URLs, and we apologize in advance if you hit a dead end; all the links were accurate as of March 31, 2006.

We hope you dive into the ocean of Imponderables presented here and plunder freely. If you feel like you are drowning in Imponderables, we'll try to throw you a life preserver—the back of this book will tell you how to get in touch with us. Until we meet again, may you always have some swash in your buckle: Ahoy, matey!

Why Did Pirates Love Parrots?

Our image of the colorful parrot astride the peg-legged, patch-eyed pirate might come from cartoons and comic strips, but the inspiration was surely Robert Louis Stevenson's *Treasure Island*, published in 1883. The beloved pet of cook Long John Silver, the parrot squawks "Pieces of eight!" with annoying regularity, and becomes the "watchbird" for the pirates after the miscreants take over the treasure hunters' fort on the island.

Stevenson admitted that he borrowed the idea of the parrot from Daniel Defoe's *Robinson Crusoe*. After being stranded on the island, Crusoe knocks a young parrot out of a tree. He teaches the bird to speak its own name ("Poll"), "the first word I ever heard spoken in the island by any mouth but my own."

But did pirates really carry parrots on their ships in real life? The evidence suggests yes. Kenneth J. Kinkor, director of project research at the Expedition Whydah Sea-Lab and Learning Center in Provincetown, Massachusetts, told *Imponderables* that "Many pirates kept parrots and other animals, as many sailors did." Kinkor says that parrots were most common among the Central American pirates who spent some time ashore, logging in places not under direct control of Spain, such as Belize, that possessed large parrot populations.

David Cordingly, former curator of the National Maritime Museum in Greenwich, England wrote in his book, *Under the Black Flag*,

> It was common for seamen who traveled in the tropics to bring back birds and animals as souvenirs of their travels. Parrots were particularly popular because they were colorful, they could be taught to speak, and they were easier to look after on board ship than monkeys and other wild animals.

Call us cynical, but pirates never struck us as the most sentimental of men. Perhaps some parrots were kept onboard as pets or mascots, but might there have been other, less humanitarian considerations? In the most-often cited contemporaneous account of the pirate world, Captain William Dampier's *A Voyage Around the World*, written in 1697, Dampier claims that his band of privateers (pirates who are authorized by a country to commandeer ships sailing other states' flags) ate parrots along with other birds, while cruising off of Venezuela.

No pirate would get fat from eating parrots, so our bet is that the primary purpose of carrying parrots was financial. In his research on pirates, David Cordingly found government records from Elizabethan times indicating that pirates gave parrots to well-placed employees of government officials, presumably as bribes.

But other folks were willing to put down hard cash to buy parrots. Dampier discusses his shipmates buying "an abundance" of cockatoos and parakeets, presumably to sell or trade. Pirates had a ready venue to sell their booty, for there were established bird markets in London and Paris in the eighteenth century, and exotic birds from the New World presumably were attractive purchases for the wealthy and status-seeking. Indeed, Cissie Fairchild wrote an entire book, *Elephant Slaves and Pampered Parrots: Exotic Animals in Eighteenth-Century Paris,* about the bustling trade in exotic animals (the Sunday bird market still exists on the Île de la Cité in Paris).

Pirates might have admired the colors of parrots, been amused by their mimicking ability, and have been satiated by their succulence. But love? Only money can buy a pirate's love.

Submitted by Tina Ritchie of Oceanside, California. Thanks also to Travis Cook of Cool Ridge, West Virginia.

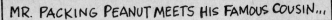

Why Do Packing Peanuts Come in Such Strange Shapes?

We have to admit ambivalence about packing peanuts—those bite-sized treats you find in shipping boxes. They do a wonderful job of protecting stuff. But they also do a fabulous job of landing all over the carpet and exhibiting static clingsmanship of the highest order.

Before packing peanuts, most shippers were content to protect merchandise with excelsior (shaved wood) or wadded-up newspapers. But a man named Arthur Graham had a better idea. Graham sold soda fountain supplies, such as ice cream cones and straws. During the manufacturing process, Graham was left with soda straw trimmings. "What could these by-products be used for?" he must have pondered.

Graham sold his "waste" to Gump's, the renowned department store in San Francisco, which had previously used excelsior as its packing material of choice. Soon, Graham had interested a defense contractor in buying much more of his straw trimmings, until the tail started wagging the dog—goodbye soda fountain business, hello wonderful world of packing materials.

The one flaw in paper as packing material is that it is relatively heavy. Graham traveled to Germany with an engineer to talk to the folks at BASF about polystyrene, and the rest is peanut history. In 1967, Graham incorporated Free-Flow Packaging (FP International), and the company still sells 100 percent recycled polystyrene "peanuts." You've undoubtedly seen FPI's peanuts—they are tube-shaped with a pinch in the middle a figure eight), betraying its paper straw ancestry (you can check them out yourself at http://www .fpintl.com/fpfp_htm).

According to Virginia Lytle of FPI, the pinch gave the peanut more stability, and the shape allowed it to interlock with its fellow eights and cushion the object sent. FPI calls its peanuts "Flo-Pak loose fill" and it is essentially the same product Graham originally invented, only with more air than the original. FPI patented the configuration of its Flo-Pak, so all of its imitators were forced to invent their own shape.

Unlike the harder foam inserts that hold electronics equipment shipped in boxes, loose fill such as peanuts is most effective when they are not rigid (most peanuts are approximately 95 percent air). Robert Krebs, of the American Chemistry Council, notes that

> the oddness of the shape gives extra cushioning and flex-
> ion as a package is put within them, and [the space in the
> cushion they make] can change shape much more easily
> when the contours are not so uniform.

We spoke to Tom Eckel, senior vice president of Storopack, the largest manufacturer of polystyrene packaging, who told *Imponderables* that every manufacturer thinks that its shape is the best— Eckel has a display board showing 35 different peanut shapes. Before committing to its choice, Storopack thoroughly researched the possibilities and hasn't changed its shape in ten years. Any contour that allows the peanuts to interlock to a certain extent but not settle into a mass is prized: the fewer the peanuts and the more air in the box, the lower the shipping costs for the sender.

Two more mini-Imponderables: If packing peanuts are made out of 100% recycled polystyrene, can you recycle yours? In the United States, at least, the Plastic Loose Fill Council will be more than happy to take them off your hands. Just call The Peanut Hotline at (800) 828-2214 and they'll direct you to a business in your community that will take them.

And even more pressing, when shipping their wares around the country, do packing peanut manufacturers protect the valuable merchandise by packing them in, er, packing peanuts?

Submitted by Eric Weiner of Pittsburgh, Pennsylvania.

Why Can't You Buy Grape Ice Cream?

Grape is a popular flavor for juice and candy—why not ice cream? Last we heard, wine was rather popular, too. When faced with this Imponderable, our sources buckled like a cheap umbrella. Carol Christison, executive director of the International Dairy-Deli-Bakery Association, wrote:

> You can't buy it because no one makes it.
> OK, OK, I really don't know. I'm sure there are some ice cream associations that might know. They make garlic ice cream, so why not grape?

Why not, indeed? It's true that we found no groundswell of outrage among consumers that grape ice cream isn't available (while pink bubblegum and licorice are!). Indeed, a representative from

Häagen-Dazs that we contacted claimed there was "no demand from consumers" for it.

The more we pursued the problem with ice cream makers, the more they decried the lack of flavor in grapes. A flavor guru from Ben & Jerry's told us

> One of the reasons we've never done a grape ice cream is because there would hardly be any taste to it. While grapes are a tasty snack, they are mostly water and don't pack a whole lot of flavor. Most of the "grape" flavored snacks that we enjoy are artificially flavored to be much stronger than what natural grape flavoring would provide.
>
> In actuality, we do use grape juice concentrate as one of the main ingredients for our line of Body & Soul ice creams [Ben & Jerry's low-carbohydrate, low-fat line of ice cream and yogurts] because the juice gives us a natural way to reduce the sugar added in.

If you look at the nutritional label of many fruit juices, especially berry juices, you'll see grape juice listed as the first ingredient. Grape juice is prized precisely because of its lack of flavor—it is used to sweeten cranberry juice, for example, but is mild enough not to distract from the cranberry taste. By using the cheaper grape juice, marketers can also trumpet "100% juice" on the labels.

Bob Graeter, the vice president of operations at what many dessert fanatics consider to be a premier American ice cream maker, Cincinnati's Graeter's, told *Imponderables* a similar tale of woe about the grape's lack of flavor. One of Graeter's franchisees once made a moderately popular grape sorbet, but it was artificially flavored, and that may be the only way to get enough grape flavor to get a good flavored sorbet or ice cream. This may be an issue as to why no one else is doing a grape-flavored ice cream or sorbet product. Our direction is to use all natural flavors in our ice cream products.

Although we have not been able to find a single grape ice cream specimen, we have found scattered sightings of grape frozen confections. Of course, grape Popsicles have been a staple for almost 100 years. Nestlé Ice Cream sells a grape sherbet as part of its Push-Up line. And Baskin-Robbins told us they occasionally offer a "purple grape-flavored ice." What do all of these frozen grape products have in common? None of the desserts except for the Nestlé Push-Up contains any dairy products, and sherbet contains only a little.

If someone provided a great grape ice cream, demand would follow. Ultimately, we think that the bad karma between grape and dairy is the villain here. We wrote to Meredith Kurtzman, who makes extraordinary frozen desserts. Although her repertoire includes exotica such as her signature olive oil gelato at New York's Otto restaurant, and more to the point, a mean Concord grape sorbet, "What's the deal?" we asked her. "You can make great ice cream out of olive oil and rice flavors but not grape?" Kurtzman responded:

Grape ice cream is possible to make, but I don't think that the flavor of grapes is strong enough to shine through a milk base. When one is making any ice cream from fruit, you have to avoid too much water present in fruit, which will make the product more icy then you might like. But then you can't add too many solids either, because they stiffen up the texture.

There are many ways to compensate for both of these factors, but that's a chapter in itself, that I'm not ready to write . . . yet.

Yet? Yet? Do I sense a cliffhanger? Will Kurtzman eliminate this Imponderable single-handedly? Hang on.

Submitted by Michael Johnson of Portsmouth, New Hampshire.

STOP THE PRESSES!

We have found grape ice cream. For more than thirty years, Tony's Ice Cream in Gastonia, North Carolina (just outside of Charlotte) has been selling grape ice cream. We spoke to patriarch Louis Coletta, the third-generation proprietor (whose children now work at the store), who told us the story of his shop and his purple ice cream.

Louis's grandfather started selling his handmade ice cream from a push cart, one bucket at a time. As he prospered, he switched to horse-drawn carts and eventually motorized trucks. The current store opened in 1947, but the object of our story, grape ice cream, wasn't introduced until the 1970s—Coletta doesn't remember the exact date of his momentous contribution to ice cream history.

When his father's health was failing, Louis, an industrial engineer, came back to run the family business. Coletta remembers grapevines always growing around his family's house, and told *Imponderables* that the reasons for offering the new flavor were twofold—it was a tribute to the family's Italian heritage, and it was designed to appeal to kids.

Coletta confirms all the problems with creating grape ice cream that our other experts detailed. Tony's has experimented using real grapes, but found that seeds were a problem, the grapes with good flavor were too sour, and the texture was wrong. So although all of

Tony's ice cream is made at a plant next door to the shop, using natural ingredients whenever possible, the only solution to the problem was to use artificial grape flavor.

Grape has proven to be one of Tony's most popular flavors, not only in cups and cones, but in milk shakes. As the family anticipated, grape ice cream is most popular with kids, but Coletta reports that another demographic group goes bonkers for grape ice cream, too—pregnant women. Want to see what drives Gastonians wild? Check out the grape ice cream photos at http://www.imponderables .com/gastonia.php.

P.S. Tony's is not alone. James Bristow, a resident of Charlotte, North Carolina, who photographed Tony's grape ice cream for our Web site, was nonplussed when contacted about our exciting find. "There's another place with grape ice cream a few miles away—Bruster's."

Right he was. Bruster's Ice Cream is a rapidly growing chain of more than 200 ice cream stores. Started as one family-run store in 1989 by Bruce Reed, Bruster's introduced grape ice cream shortly after it opened. According to Reed, kids requested the flavor. We spoke to Christina Parker, vice president of operations, who told us that Bruster's grape ice cream was also made from artificial flavoring, and that the flavor is still popular among children, especially in the summer. And if grape ice cream seems a little mundane for the kiddies, they can also sample other flavors such as blue pop, watermelon, purple dinosaur, cotton candy explosion (with chocolate-covered Pop Rocks), and our favorite, birthday cake (with icing and sprinkles, and presumably no candles).

Why Don't You Ever Hear Giraffes Vocalize?
Do They Ever Make a Sound?

Giraffes are among the most taciturn of animals. We've never heard a giraffe vocalize but it turns out that they've been dissing us, for they possess larynxes and vocal cords and actually make a variety of sounds.

The heroine of our story is Elizabeth von Muggenthaler, a bioacoustician and president of Fauna Communications Research Institute in Hillsborough, North Carolina. Muggenthaler reckoned that because giraffes are highly social, and are forager-vegetarians who are the prey of other animals, it was highly unlikely that they could survive without intraspecies communication. The giraffe's anatomy was another clue to Muggenthaler—if they don't speak, then why are their ears shaped like parabolas, which seem perfectly designed to pick up on sounds?

Giraffes do vocalize occasionally. Calves, especially, utter a bleating mewl. Mothers utter a "roaring bellow" when looking for their young, who are often left alone in the forest while the parent forages. And males are known to seduce partners with a "raucous cough." Occasionally, adults also bleat (Muggenthaler compares them to goat bleats) and "moo" as if imitating cows. When threatened, the giraffe's yelling side emerges—they are capable of mustering up a roar when in danger. Still, these vocalizations are exceptions rather than the rule, and it is possible even for those who work around giraffes to think they are mute.

Could giraffes speak in ways humans couldn't understand? Scientists had already determined that whales and elephants communicated via infrasonic sound—vocalizations at such low frequencies that humans could not hear them. Scientists discovered the songs of the humpback whale more than forty years ago, and researchers like Muggenthaler and the Bioacoustics Research Program at the Cornell Lab of Ornithology have documented the complexity of infrasonic vocalizations. A Cornell researcher, Katy Payne, discovered the elephants' infrasonic communication when she found her ears throbbing near elephant cages. It reminded her of singing in a church choir, where the pipe organ was almost inaudible at the lower frequencies but the pressure in her ears palpable.

While studying the low-frequency vocalizations of elephants, Muggenthaler discovered that rhinoceroses also utilized infrasonic vocalizations, and she suspected that giraffes did, too. In 1998, she confirmed it by studying eleven giraffes in two zoos. Measuring infrasonic communication is difficult in the controlled atmosphere of a zoo, since passing cars, ambient wind, and even water create infrasound, but it is even more difficult in the wild, where other animals also can compete with the giraffes' vocalizations.

Muggenthaler and her fellow researchers discovered that the

giraffes' infrasonic vocalizations were associated with two physical movements: a "neck stretch," when giraffes lift their head and necks over their bodies, and the "head throw," that features a lowering and quick raising of the chin. Almost every time a giraffe was observed performing a neck stretch, an infrasonic vocalization accompanied it. Head throws were more common, but there giraffes vocalized only 25 percent of the time.

Although it hasn't yet been proven, Muggenthaler's theory is that the infrasonic vocalizations might be caused by

> large volumes of air being forced up the neck and/or possi-
> bly channeled through hollow posterior sinuses. During the
> study, observers noticed a "shiver" or vibration extending
> from the chest up the entire length of the trachea that oc-
> curred during some neck stretches that accompanied vocaliza-
> tions. It is possible that this "shiver" is air movement, and
> could be responsible for the signal. If air is [sic] moving up
> the giraffe's neck is producing infrasound, the mechanism
> may be Helmholtz resonance, which occurs when an enclosed
> volume of air is coupled to the outside free air by means of an
> aperture.

If giraffes are capable of vocalizing in a higher frequency through their mouths, why bother with the low-pitched stuff? One obvious advantage is that low-frequency sounds can travel farther than higher-pitched ones, a crucial advantage to giraffes (and elephants), who often are separated from their families by greater distances than their voices can reach. And although Muggenthaler's team did not study how the giraffes use infrasonic vocalizations to communicate with each other, she does speculate about why infrasonic communication might aid in giraffes' survival. Evidently, we are not the only animals who can't hear their low-frequency emissions:

> If the giraffes are communicating [with each other], it would be very advantageous for them, being prey, to be able to communicate "covertly" using signals designed to blend in with the background noise.

Submitted by Peter Lanza of Stamford, Connecticut.

Why Do Some Slot Machines Use Fruit Symbols?

When you think about slot machines, chances are you conjure up glassy-eyed gamblers in Sin City, Nevada, with cigarettes dangling from their mouths—hardly the setting for showcasing images of fresh fruit. Yet fruit was associated with slot machines almost from the time of their invention—in fact, in England, slot machines have always been known as "fruit machines."

Although there were mechanical gambling devices before, including a primitive precursor of today's video poker machines, Charles Fey invented the first one-armed bandit in 1895. Fey's Liberty Bell slot machine, with three reels sporting three of the four suits found in a deck of cards (clubs were the odd suit out) and the now-familiar cracked

Liberty Bell. The highest jackpot, the princely sum of ten nickels, was won if you could line up three Liberty Bells in a row.

Edibles came into the picture when the Mills Company of Chicago redesigned Fey's original Liberty Bell and created a special machine for the Bell Fruit Gum Company. While most early slot machines were gambling devices placed in taverns (prizes were often a free drink or small amounts of money), Bell wanted a machine that could be played for a family audience at fairs and carnivals. Instead of playing cards, Bell placed drawings of fruits that represented the flavors of Bell Gum. If three watermelons or lemons were aligned, the machine would dispense a pack of Bell Gum. With the great popularity of this machine, the fruit symbols prevailed, and are still depicted on some modern machines.

And if we may throw in our own mini-Imponderable, we always wondered what the bar symbol on slot machines signified. We assumed they were meant to be gold bars, but they weren't tapered like them. It turns out that the bar was a stylized version of Bell Fruit Gum's logo, now an example of obsolete product placement.

With the advent of nine-line slots with themes ranging from Monopoly to *I Dream of Jeannie,* the fruit symbols are a withering but not yet dead symbol of old-school gambling. It's hard for a lemon to compete with Elvis- or *Star Wars*–themed slot machines for a gambler's attention.

Submitted by Faye Railing of San Diego, California.

Why Does Lightning Have
a Zigzag Pattern?

We're always pleased to meet a source who is enthusiastic about his work. Matt Bragaw, the lightning specialist at the National Weather Service Forecast Office in Melbourne, Florida, is such a guy. He shares his passion about lightning on his corner of his office's Web site (at http://www.srh.noaa.gov/mlb/ltgcenter/whatis.html, including a nifty animation of a lightning strike). Matt was kind enough to answer some of our incessant follow-up questions. He warned us that although lightning was one of the earliest remarked upon natural phenomena, it is one of the least understood, with many of the major discoveries about it having been made in only the past fifteen years.

Although there are other kinds of lightning, such as heat lightning and Saint Elmo's fire, the familiar zigzag lightning we're talking about here is cloud-to-ground lightning (lightning inside a cloud, also known as cloud-to-cloud lightning, is actually more prevalent). Before we see any sign of lightning on the ground, turbulent wind conditions send water droplets up the cloud while ice particles fall downward. The top of the cloud usually carries a strong positive charge and the bottom a negative one. During the movement of the ice and water droplets within the cloud, electrons shear off the rising droplets and stick to the falling ice crystals. The opposite charges attract until a tremendous electrical charge occurs within the cloud. When the cloud can no longer hold the electrical field, sometimes a faint, negatively charged ladder channel, called the "stepped leader," materializes from the bottom of the cloud.

While it might appear to us as if the bolt of lightning strikes the earth instantaneously, in one zigzag strike, what you are actually seeing is a whole series of steps, which are only about 50 meters in length. In an e-mail to *Imponderables,* Bragaw elaborates:

> In what can be described as an "avalanche of electrons," the leader's path often splits, resplits, and re-resplits, eventually taking on a tendril-like appearance. Between each step, there is a pause of about fifty microseconds, during which time the stepped leader "looks" for an object to strike. If none is "seen," it takes another step, "looks" for something to strike, etc. This process is repeated until the leader "finds" a target.
>
> *It is this "stepped" process that gives lightning its jagged appearance* . . . Studies of individual strikes have shown a single leader can be comprised of more than 10,000 steps!"

Once the leader hits the ground, all of the other branches of the stepped leader's channel stop propagation toward the earth.

We mentioned that the stepped leader is faint as it leaves the cloud and heads toward the ground. If so, then why is lightning usually so bright? The negatively charged stepped leader repels all negative charge in the ground, while attracting all positive charge, which sends energy back from the ground to the clouds. This "return stroke" occurs in less than 100 microseconds, which is why we can't differentiate cloud-to-ground movement from ground-to-air. But this upward process, according to Bragaw, "produces almost all the luminosity" that we see when we think we are observing cloud-to-ground lightning strikes.

Twenty to fifty milliseconds (thousandths of a second) after the initial return stroke stops flowing up the channel, "leftover" electrical energy in the cloud often sends more leaders down to the earth in the same channel. Because these "dart leaders" use an already-established channel, they discharge continuously instead of in steps. Even though these subsequent dart leaders don't need to stop to look for places to hit, as their route is the same as for the first leader, you'll still see the familiar zigzag. As Bragaw puts it:

> Because the stepped leader initially burns a jagged path, all lightning takes on a jagged appearance.

Submitted by Robert Underwood of Blue Ridge, Virginia.

DAVID FELDMAN

Why Do Men Have Domain Over the Family Barbecue?

The allure of the grill compels men. Even guys who refuse to clear the dishes or toss a salad often spend hours in ritualistic trances, tending to their "Q." Button-downed types willingly don aprons ladened with inane jokes.

According to the Barbecue Industry Association, 84 percent of Americans own some sort of barbecue or grill (charcoal or gas). Sixty-one percent of men barbecue periodically, while only 39 percent of women participate in the thrill of the grill. Weber-Stephens, the largest manufacturer of outdoor grills, commissioned their own survey in 1999 and found an even greater disparity: 63 percent of the grilling was performed by men, 22 percent by women, and by both in 15 percent of the households that barbecued regularly. Barbecue grills are used both to grill (cooking directly over high heat) and to barbecue (slow-cooking over indirect heat). Just as barbecue industry statistics usually combine both cooking methods under the rubric, "barbecue," so are some of the sources quoted below referring to barbecuing when they are actually grilling.

Barbecuing no doubt dates back to the cavemen. It's all well and good to bash a critter over the head with a club, but how do you preserve that big hunk of meat? Before refrigeration, nearby heating agents (trees) were used as the fuel to cook and then preserve freshly killed animals.

When the early Spanish settlers came to the New World and found natives using smoky fires as a way of preserving large slabs of meat in the sun (and keeping insects away), they quickly adopted these methods as they settled in the southwestern United States. Well before the Civil War, especially in the South, barbecues became a favorite form of cooking at large parties and celebrations. Pits were dug in the ground and filled with wood that burned all night to create a bed of coals. Whole carcasses of pigs or cows were hung from rods or laid to cook on grills above the coals. And those carcasses had to be turned above all that sooty wood. This work, requiring strength and a high tolerance for heat and grime, was performed by "pit men," usually black slaves.

Barbecues became a fixture at political rallies, too. Who couldn't draw a crowd with the lure of free barbecued meat, lemonade, and perhaps some whiskey, too? Without the problem of having to cater to females (who couldn't vote), barbecue rallies consisted of male politicians providing food cooked by males for a male voting constituency.

After the Civil War, in the old West, cowboys often cooked whole cows on a spit over the campfire. Today we see men huddling around the home grill, recreating the cowboy's campfire culinary tradition, without the soot or the horses or the heavy lifting.

The earliest barbecue restaurants were easy to start. Some "pit men," full-time farmers during the week, opened up de facto restaurants, usually little more than shacks, on the weekend. This tradition continues to this day, when some of the best barbecue in many cities is provided by unlicensed "amateurs," who tend pits on the weekend.

Even now, cooking a whole pig or side of beef is hardly a glamorous undertaking, but grilling hot dogs and precut chicken pieces is more our speed. There isn't much need for testosterone to tend the family barbecue. Our portable grills are relatively pristine. Worried about the heat and unpredictability of a wood fire? Then substitute our modern equivalents: charcoal briquettes. If briquettes are too dirty or difficult to light, you can switch to the even more convenient gas grill, which is no harder to control than the indoor range.

If the modern barbecue is so much easier for women to handle, the historical antecedents can't explain why men are still eager to wield the tongs. We contacted scores of amateur barbecue enthusiasts, some of them participants on the competitive barbecue circuit, and found four dominant theories for males hogging the barbecues:

1. *The Great Outdoors*

We were surprised how often men mentioned being outside as one of the great appeals of barbecuing, and how the grill reinforces the traditional roles of men and women in the household (in many families, the wife maintains the inside of the house and the husband performs the outdoor chores). As one anonymous griller e-mailed us:

> By the time Dad gets home from work, Mom is more than willing to let the kids flock to Dad around the barbecue pit. Mom enjoys the solitude of the empty house to freshen up

and recharge. The kids love how playful and relaxed Dad gets after he crunches his third empty beer can.

Even if it's only on a patio, any outdoor setting is great if there is a barbecue involved, according to the devotees we consulted. Some think that the appeal of outdoor cooking hearkens back to our ancestors, and is a tribute, of sorts, to their self-reliance. As barbecue aficionado Chris Bennett of New Bern, North Carolina, puts it:

> I think that a large part of the thrill of it goes back to chuck-wagon days or even prehistory. There is something very satisfying about taming the flame and cooking in this somewhat primitive manner. I remember that early in my grilling days, I took almost absurd pleasure in preparing a meal that came almost entirely from my own labors. I cooked the meat, grew the vegetables, made the barbecue sauce, and drank my own homebrewed beer. The only part not of my labors was the meat.
>
> I don't hunt. But since I work, perhaps this too came from my labors. The pleasure came from the idea that I was providing for my family and showing love through my efforts.
>
> I asked my wife this question and she said that men are idiots and don't have the sense to stay in the air-conditioned house and cook. She might have something there.

2. Male Bonding

Chris Bennett also observes:

> Another aspect of grilling that deserves mention is the camaraderie it engenders. The grill gives men a place to stand and talk and renew acquaintance, while the smoke chases the

women away. I hope this does not sound too sexist—it isn't meant that way.

Don't worry, Chris. Your wife deserves payback for that "idiot" crack.

Not only do men tend to barbecue as a way of hanging out with their friends, but if they play their cards properly, they can earn brownie points, too. Willy, a grilling fanatic, noted that perhaps barbecuing links modern man with his caveman heritage, but he has a new theory, which he calls the "golfing theory." Willy barbecues and plays golf with the same friends. Somehow, his wife doesn't appreciate him going out three days a week to golf and drink beer with his buddies:

> Instead, we say, "Honey, I'm going out with the boys and we're going to cook some ribs and butt. It should be done by 6:00 p.m." Suddenly, you're helping out. You're still drinking and shooting the breeze with the boys, but now it's culturally all right and you may even be a sensitive guy.

3. Sex Roles Rule

Pat Nicholas, who with her husband is a regular on the Texas competitive barbecue circuit, wrote *Imponderables* that most women prefer avoiding dirty charcoal and usually have to provide the fixings that go with barbecue anyway. Have these gender stereotypes been reinforced by the mass media? How often have you ever seen a fictional depiction or advertisement portraying a woman barbecuing? Nowadays, approximately 20 percent of competitive grillers are female, a huge jump from decades ago.

Yet comedienne Rita Rudner's observation about men's affection for barbecuing is pertinent: "Men will cook if danger is involved." One of the attractions of the grill to some men is that barbecue equipment is not delicate. Compared to a stove or oven, the barbecue grill is like a power tool or a pickup truck. Many men have a strange predilection for cars; barbecues share many similarities, including wheels, generation of smoke, and, in a crowd, no shortage of backseat drivers. With both the outdoor grill and the car, one has to look under the hood when trouble is afoot, even if you have no idea how to solve the problem.

So our culture has deemed barbecuing a "male thing," and for some, letting a woman barbecue would seem "unmanly," especially to the neighbors. On the animated television show, *King of the Hill*, the Texas males determine social pecking order by the condition of their lawns. Particularly in the Sun Belt, this peer competition often exists, and it is the males who are held responsible for the lawn. Bill MacKenzie, vice president of the Greater Omaha BBQ Society, maintains that:

> Perhaps there is a social pressure or expectation put on men to be barbecue chefs, a sort of "keeping up with the Joneses" that starts with the yard and home exterior and would seem to have an extension onto the patio and outdoor grill.

4. A Hobby, Not a Chore
Derrick Riches, About.com's barbecue guide (http://bbq .about.com) wrote us:

> The day-to-day cooking traditionally carried out by women has always been a chore, something that needed to be done and quickly grew to be a somewhat undesirable task. Backyard cooking is more like a hobby. You need special equipment and the more you practice the better you get. I think this is another draw to men to do the grilling.

Although some married male grillers enjoy preparing the side dishes, our correspondents report that it is usually the wife who issues the invitations, prepares the non–barbecue-related food, sets the table, and cleans the dishes. Some women, resigned to their fate, have decided to go with the flow. Even if she can't get her husband to agree to food preparation parity, this barbecue widow, who prefers to remain anonymous, will take what she can get:

> If I were the one on whom the family depended to plan, prepare, and put most of the meals on the table every single day, all year long, and if, on occasion, someone else said: "Hey honey, don't worry about dinner tonight, I've got it covered," I'd be up in a Calgon tub with candles, a book, and a beverage until I was called down to the table (which I probably set myself). I would *not* be hovering anywhere near the food preparer, even if he was having a weekend blast out at the Q pit. Hey, I caught a break from routine here—why should I?

More and more women seem to be grilling, though. Most of the barbecue experts we spoke to attributed the rise to the popularity of gas grills, which don't require starting fires and are much cleaner than their charcoal counterparts. Increasingly families are experimenting with grilling and smoking fish and vegetables, which have more feminine appeal.

Will as many women as men barbecue fifty years from now? Our guess is no, at least for traditional wood and charcoal Q. As long as there are unmanageable fires and grease and grime, our guess is that men will still consider barbecue to be their domain, and women will tend to "leave it to the boys." One of our correspondents, when posed this Imponderable, indicated that barbecuing does wonders for fidelity in his marriage, even if un-intentionally:

> The wife doesn't have to worry about other women coming
> around, especially when the husband smells of beer, wood
> smoke, and meat drippings.

Submitted by Ronald Walker of El Segundo, California. Thanks also to Cindy McDonald of San Francisco, California; Ethan Jennings of Dover, New Hampshire; and Tracy Takach, via the Internet.

Why Do Bats Roost Upside Down? What Prevents Them From Falling Down?

Bats are a tad eccentric. They are the only mammals that can fly. At night, they flutter around, snarfing up assorted bugs for food; and during the day, most species of bats literally hang out: upside down.

The key ingredient in allowing bats to roost upside down is their specialized musculature. When humans try to grip something with our hands, say, hanging on a horizontal bar, we clench our muscles, straining to keep not only our hands and wrists locked, but aggravating our shoulders and arms. It's almost the exact opposite for bats. When a bat finds a suitable roosting site, it opens its claws and grabs with its talons; it doesn't clench its muscles, but rather relaxes them. The weight of the upper body actually keeps the talons locked, so it takes no more exertion for a bat to roost than for us to recline on a Barcalounger. The lack of effort needed to stay roosting allows bats to enjoy a form of hibernation known as "torpor." During torpor, which bats can induce at will, their body temperature and blood pressure decrease, and they barely move. In very cold weather, bats can enter a full hibernation mode, roosting blissfully upside down the entire time.

How little effort does it take a bat to roost upside down? Dead bats are routinely found in typical roosting position, looking like they are just taking a snooze. Only when they want to take off from their roost do bats have to flex their muscles.

Other anatomical oddities also help bats roost upside down. Their necks are extremely flexible, so if they need to look behind them—no problem—they can turn their heads 180 degrees. The hind legs of bats are rotated so that their knees face backward, which aids in roosting. Most bats don't have the ability to give birth upside down, though, so most species literally hang by their thumbs while delivering (as they do when urinating and defecating), which does take exertion.

So we know *how* bats roost upside down; now let's look at the *why*.

1. Bats' legs are weak. In exchange for the unique ability of bats to fly, Mother Nature saddled them with unusually weak legs, with light and slender bones. Light legs allow bats to fly faster, but lower their ability to stand, walk, and support their own weight so that they could perch like birds.

2. Roosting upside down improves bats' takeoffs. Although bats are efficient flying machines once aloft, their wings aren't strong enough to enable them to take off from the ground the way birds can. If bats are attacked by a predator, their roosting position allows them to escape quickly, even if they are in torpor, by simply dropping off their roosting spot. Some bat researchers believe that at one time in their evolution, bats might have been gliders rather than flyers, as their takeoff pattern suits an animal unable to fly.

3. Predators are foiled. Bats' predators include owls, hawks, snakes, raccoons, and in many places, humans. Bats evade many of their predators simply by being active when their enemies are asleep. But roosting upside down allows bats the opportunity to find roosting sites unappealing to or unreachable by predators, such as the roofs of caves or the ceilings of attics and barns.

Submitted by David O'Connor of Willoughby, Ohio. Thanks also to Renee Gonsiewski of Villa Park, Illinois; Michael Wille of Springhill, Florida; Michael Cipoletta, Jr. of Malden, Massachusetts; Sharri Browne of Fort Nelson, British Columbia; and Christian Morrow of Santa Barbara, California.

Why Are Most Bibles Printed in Two Columns?

As a clergyman is one of the three readers who asked this Imponderable, we had a sneaking suspicion that the answer lies not in religious conviction but in convenience. Although it contains only what later became five books of the Old Testament, the first iteration of the Bible, the Jewish Torah, presented enormous logistical difficulties to produce. The Torah was hand-scribed on sheepskin scrolls; if unfurled, one measured approximately 150 feet long. The earliest Bibles, written in ancient Hebrew and Greek, were all in scrolls. Imagine the logistical nightmare if each line ran the total length of the scroll—thus columns were a necessity.

The use of columns continued in codices, the earliest books. According to Erroll F. Rhodes, biblical scholar and translator of

The Text of the Old Testament: An Introduction to the Biblia Hebraica, the important surviving biblical manuscripts from the fourth to the seventh centuries **a.d.** range in format from a single column to four columns. The first English-language Bible, hand-written in English in the 1380s by John Wycliffe, also used a two-column format.

But perhaps the most influential historical precedent was the first printed Bible, created by Johann Gutenberg in 1455. The Latin text was first presented in pages of two columns with forty lines per page (although this was quickly changed to forty-two lines in later printings). Other printers followed suit, as Rhodes described to *Imponderables:*

> The first printed English Bible (1535) continued the tradition, and the adoption of the King James Version of 1611 probably led to the popularity the format enjoys today. Economy of space may have played some role, but the traditional association of form with content was undoubtedly a factor for many, both publishers and readers, to whom the Bible was primarily an icon for traditional stability and values.

The two-column format is adopted by many other printed works, especially reference books such as encyclopedias and dictionaries. Kang-Yup Na, a religion professor at Westminster College in New Wilmington, Pennsylvania, thought that the answer to this question lies more with readability than tradition or faith:

> I have a hunch that it has more to do with facilitating a clearer and faster reading of the text. The width that the eyes can scan quickly cannot be much wider than what you see in *Time* magazine—at least not for the average reader. The

columns allow for a more accurate reading because you're less likely to skip or repeat a line.

Or as one biblical expert, who asked us to remain anonymous, put it a tad more bluntly:

This question is not scholarly but rather a trade question.

Submitted by Mrs. Elmer Neumann of Granite City, Illinois. Thanks also to Gregg Hoover of Pueblo, California; and Reverend Duane Breaux of Pierre Part, Louisiana.

Why Do Bottles Have Necks?

Bottles date back to the ancient civilizations of Egypt and Mesopotamia. Historians think that wine consumption started there as far back as 5400 b.c., but their bottles didn't resemble ours at all—they were amphorae, clay flasks with necks shorter than an NFL lineman's. Relics indicate amphorae were stoppered with cloth, pieces of leather, or fired clay.

When glassblowing was developed during Roman times, most of the vessels were squat and onion-shaped, probably because these were easier to manufacture. This shape was badly suited for wine-makers, in particular, because wine bottles need to be stored on their sides, in order for the wine to stay in contact with the cork (dry corks crumble and may allow air to enter the bottle—oxygen is the enemy of aging wine properly).

Before mass production of glass bottles, there was no uniformity in size, but by the nineteenth century, most wine bottles were 700 to 800ml, with 750ml not becoming the standard until well into the twentieth century. The long-neck glass bottle has only been with us for a couple hundred years, so we were a little surprised when an authority like renowned bottle collector Lieutenant Commander J. Carl Sturm replied to our query with: "This question has never been raised that I know of in my 44 years of collecting old bottles. I can only theorize as I know of nothing in print."

We spoke to or corresponded with nearly thirty experts on bottles—bottle collectors, glassblowers, bottle manufacturers, and winery executives—and we were offered plenty of theories but no smoking-gun answers. We couldn't even gather a definitive conclusion about whether the longish neck was a byproduct of the glassblowing process, an aesthetic decision, or a utilitarian feature.

Steve Fulkerson, general manager of Saint-Gobain Containers, a company that makes many different kinds of bottles, including bottles for Simi Winery, advocates the by-product theory:

> Early glass containers were hand blown by inserting a tube
> into a gob of molten glass and blowing on the tube by mouth.
> As the bubble expanded the outside of the bubble was shaped
> by the use of a wooden paddle. A container with a neck was
> easier to handle as the inside of the neck surrounded the last
> few inches of the tube, making it less likely that the gob
> would fall from the tube as it was being shaped.

Rick Baldwin, Midwest regional director of the Federation of Historical Bottle Collectors, adds that:

> A neck, the end of which was tooled to form a heavier
> lip, afforded a smaller area for a closure/seal. A smaller area

DAVID FELDMAN

would be easier for the glassblower to form a nice, "perfectly circular" sealable lip, and the task could be done faster.

But just as many experts felt that the neck had nothing to do with glassblowing methods. Sturm is in this school:

> There are quite a few containers with wide mouths that were used for food storage, etc. The glass blower had the knowledge and capability of cutting the container so that it didn't have a neck.

The glassblowers we contacted agreed, and pointed out the same design advantages of necks that most of the bottle historians noted:

1. Bottles with necks are easier and cheaper to seal. Fred Holabird, historical consultant and owner of Holabird Americana auction house, told *Imponderables* that

> My guess is that it's cheaper to have a stopper that's small. If you put the equivalent of a mason jar lid in a wide bottle, the cost is proportionally higher. So it's cheaper to make an enclosure with a small lid.

John Pritchard, operations manager of Simi Winery, notes that the neck of the bottle has always been used to hold the cork. Necks have been "pretty much" standardized now to a length of two and one-half inches in order to hold a two-inch cork effectively. The narrow neck decreased evaporation (important when early bottles were used to hold expensive perfume oils) and exposure to air.

2. *Necks provide a handle*. Rick Baldwin, who notes that "you've already caused me to lose three nights' sleep pondering the question," observes that long necks were long ago applied to large, bulbous flasks "so that they could be drunk from directly." Paul Bates, who with son Tom established the Museum of Beverage Containers and Advertising in Millersville, Tennessee, points out a practical advantage to the long neck that was particularly true in the past—in the early days, bottles were much heavier than they are now. Without a neck, these bottles would have been more likely to slip out of one's hands. And for wines, in particular, the neck acts as a handle when grabbing a bottle stored horizontally.

3. *Necks facilitate pouring*. Traffic jams are referred to as "bottlenecks" for a reason. Rick Baldwin argues that a neck, and its smaller diameter opening, affords a slower and more restricted flow of contents, whether one is decanting a liquid into another vessel or swigging it straight from the bottle.

Bottle manufacturer Steve Fulkerson told *Imponderables* that the neck plays an important part in the "pourability" of a bottle. He dared us to try to pour wine from a wide mouth jar into a wine glass (no thanks, we have enough problems pouring from a bottle).

The pourability is a key psychological component in the customer's appreciation of a bottle and its contents. Fulkerson once had a customer who asked if his company could manufacture a ketchup bottle that allowed the ketchup to pour more smoothly. The customer noted Heinz's success with boasting about the slow progress of its ketchup from bottle to plate, and wanted his thinner product "to pour with the same 'anticipation' as Heinz."

Despite attempts by beer marketers to push stubby alternatives, we seem to have a love connection with swanlike long-necks, whether we are pouring wine, beer, or ketchup.

Submitted by Matt, via the Internet.

1962 AL ROGERS 1963

Career wins: 0 Career

AL "Line Drive" ROGERS

#14 · PITCHER

Shown in his famous ↑
post-pitch position ↑

HENDERSON Moles

Why Is the Pitcher's Mound Elevated on a Baseball Diamond? And Come to Think of It, Why Is There a Pitcher's Mound in the First Place?

Pitcher's mounds were not mentioned in the official baseball rules until 1903, when Rule 1, Sec. 2 specified that "the pitcher's plate shall not be more than 15 inches higher than the base lines or home plate." As we discussed in *Do Penguins Have Knees?*, since the beginning of baseball, the pitcher has been moved progressively farther and farther away from the batter. As overhand pitching started to replace the underhand delivery, batters were given more and more distance from the pitcher. In the 1870s, the pitcher stood forty-five feet from home plate; in 1881, the rule was changed to fifty feet; in 1887, fifty-five and a half feet, and only in 1893 did today's weird but enduring sixty-feet, six-inch standard prevail. The farther you move back the pitcher, the more offense can be generated. Batters needed the extra space to try to figure out the velocity and movement of the ball.

An elevated mound benefits the pitcher because it provides them an artificial height advantage. Batters prefer hitting a "flat" ball coming at them—the extra downward angle from a pitch off the mound makes it harder for the batsman to make solid contact. The elevated mound also allows gravity to further assist the pitcher: given the same thrust off the pitcher's hand, a ball thrown from a higher elevation will gather more kinetic energy and a little more speed by the time it reaches the batter.

But the biggest advantage of throwing off the mound is the aid it gives to the pitcher's mechanics, as baseball historian and author Bill Deane wrote to *Imponderables*:

> Throwing a ball downhill is always easier than throwing on a level or slightly uphill. The extra momentum which the slope allowed the pitcher to generate in the downward shift of his weight and lead foot and the overhand thrust of the throwing arm [and thrust of the torso] made for increased velocity— always an advantage in baseball.

Before the 1903 rule, there was no standard for the elevation of mounds, and pitchers presumably had a hard time adjusting to varying heights and slopes of different ballparks. Bill Deane notes that photographs from the 1880s indicate that pitching mounds grew higher and higher as overhand pitching became the vogue. Fielders must have had a hard time chasing down pop flies in the vicinity of the pitcher's mound, when they didn't know, with their eyes on the ball in the sky, whether they were about to trip over a molehill or bump into a mountain. For the safety of the players and to mollify pitchers and teams upset at the tendency of some ballparks to install eccentric mounds, there was a need for the standardization which was imposed in 1903.

But the standard did not stand. In an effort to bring more offense to Major League Baseball, the pitcher's mound was reduced to the height of ten inches in 1969. The shift did coincide with an era of better hitting statistics, so the change might have served its purpose. Yet as the average height of pitchers keeps growing, today's typical 6 foot 3 inch pitcher, let alone a 6 foot 10 inch–Randy Johnson, doesn't stand much lower on the mound than his predecessor a century before.

We can't conclude, however, that the original purpose of the elevated pitcher's mound was solely, or even mostly, to aid the pitcher. Most likely, the initial impetus for the elevated mound had more to do with groundskeeping than pitching. By installing a slope on an otherwise flat field, much better drainage was afforded a busy part of the diamond when the weather turned inclement. As Deane puts it,

> A game couldn't be played if the pitcher's area was a quagmire. After rain storms or rain delays during games, it was probably the practice to throw some sand or dry soil onto the mound to absorb excess moisture. Probably the sheer accumulation of these treatments accounted for the rising elevation of the mound area.

Submitted by "Jerry," a caller on the late John Otto's show, WGR, Buffalo, New York. Thanks also to John Ryan of Portsmouth, Rhode Island.

Why Are Most Psychics Women?

Every expert we consulted agreed that the percentage of female professional psychics lies somewhere in the 75 to 90 percent range. If you include storefront fortune-tellers and palm readers, who are almost exclusively female, the percentages would go up.

In order to solve this Imponderable, at least three separate questions must be answered: Do women, in general, have more skills or qualities necessary to do this job? Is there a reason why patrons might prefer to consult a female than a male psychic? And are there economic, psychological, or sociological reasons why women might gravitate toward this job (or why men might be repelled by the notion, or not have the opportunities that women are afforded)?

In order to unravel all of these threads, we went a little nuts here at Imponderables Central, speaking to some of the most prominent psychics and skeptics, gypsies engaged in fortune-telling, and police specializing in fraud, and also reading scores of books on the subject. Not one written source addressed this Imponderable head on, but every person we contacted directly had an opinion on the subject.

QUALITIES NEEDED TO BE A PSYCHIC

Many of our sources contend that women, in general, are superior listeners and talkers compared to men. Communication skills mean little, however, if a potential professional does not possess the abilities to "read" a potential "sitter" (a client of a psychic). Is there any substance to the notion of a "woman's intuition"?

Dr. Doreen Virtue, "spiritual counselor" and author of *Goddesses and Angels* and *Divine Prescriptions,* and a certified professional counselor as well as a Ph.D. in counseling psychology, argues that there is. Virtue feels that women's roles as child bearers and, usually, child nurturers, are instrumental in their developing the skills necessary to be a psychic counselor. Virtue told *Imponderables* that women must trust their instincts about the needs and feelings of their babies at a time when the child cannot communicate desires through words; through this experience, women gain greater confidence in exploring nonverbal realms.

James Randi, president of the James Randi Educational Foundation, and perhaps the world's most prominent skeptic and psychic debunker, agrees with Virtue that women's traditional role as child nurturers has "hardwired" them to become more observant of the "little things"—emotional nuances and relationship frictions—that men might not notice. But he sees no evidence that women are superior to men in predicting the future or in mastering the paranormal; only in "reading people" more successfully. Randi's foundation offers a one million dollar prize to anyone who can prove verifiable psychic ability. So far, no woman (or man) has been able to gain the prize.

Most of the psychics we interviewed for this book endorse the view of Brian Edward Hurst, a spiritual medium from Reseda, California, who was a teacher of the bestselling author-psychic James Van Praagh, and author of *Some Go Haunting.* Hurst believes that although more women demonstrate strong intuitive skills, the few men who "get in touch with their psychic energies" are every bit as effective as women.

Psychics felt that most men are less open to the spiritual realm than women. Many studies have shown that people who believe in psychic phenomena "before the fact" are much more likely to experience it. Virtue goes so far as to say that some men would not accept as valid a psychic experience that has just happened to them because it conflicts with their belief system.

Women are not only more likely to believe in the truth of an intuitive insight, but to act on it. Perhaps the most famous psychic in the United States, Sylvia Browne (http://www.sylvia.org), told *Imponderables* that women are much more likely to act based on gut feelings that defy rational analysis. While men may have hunches, they will tend to try to back up the hunch with facts and logic, whereas Browne claims that women are more likely to trust the emotional: "Intuition—the gut feeling—is our cell phone to God."

Louise Hauck, a "clairvoyant spiritual counselor" (http://www.louisehauck.com) and author of *Beyond Boundaries: The Adventures of a Seer,* believes that the men who make the most successful psychics have been the ones who are most sensitive ("capable of getting out of their heads and into their hearts")—in essence, men who possess personality characteristics that Western culture have traditionally deemed "feminine." Frederick Woodruff (http://zenpop .home.mindspring.com/), author of *Secrets of a Telephone Psychic,* feels strongly that cultural assumptions are responsible for the relative dearth of male psychics and the disdain with which they are treated by the mainstream press. He wrote to *Imponderables:*

> Psychics, astrologers, tarot readers, intuitives—all of these belong to the archetype of the feminine, and this explains why our culture . . . has a hard time accepting the viability of intuitive arts. Sociologically speaking, feminine ways of perception and expression are not honored and promoted the same way masculine principles are upheld. Because survival and security are such compelling drives, the logical, rational, practical approach holds precedence.

Woodruff is gay, and many of the psychics we talked to mentioned that most of the male psychics they know are gay. While no one claimed there was a link between sexual preference and psychic ability, in his book Woodruff notes that even as a child,

> the driving force behind my fascination with mysticism was an early and abiding attraction to and identification with the feminine realm.

Psychologists have long confirmed the conventional wisdom that women are superior at reading body language and other nonverbal cues. Skeptics argue that the very qualities that make women more attuned to the emotional needs of others also make them better con artists than men. What good is purveying information if you won't be perceived as accurate, empathic, or honest? As one female skeptic e-mailed us: "Women are better liars."

THE CLIENTS OF PROFESSIONAL PSYCHICS

By all accounts, many more women visit psychics than men (figures most often cited are in the 75 to 90 percent range, similar to the breakdown of the psychics themselves). Gallup Polls consistently indicate that women have a stronger belief in psychic phenomena than men. Might a woman simply feel more comfortable consulting another woman about intimate details of her personal life? Does a female psychic represent a more maternal, forgiving, and nurturing figure? Skeptic W. Rory Coker, a physics professor at the University of Texas at Austin, has studied fortune-tellers and observes that women, in general, seem to feel more comfortable with seeking external help in order to solve problems; unlike men, women do not feel guilty or diminished if they lean on loved ones, friends, or counselors.

Although there don't seem to be any formal studies of whether clients of psychics prefer male or female practitioners, scores of studies about the preferences in psychotherapists have been conducted. Is it far-fetched to compare a psychic reading with a psychotherapy session? In *Secrets of a Telephone Psychic,* Frederick Woodruff muses:

> Why do people reveal so much more, so quickly, to a psychic than they would to a therapist or family counselor? My theory is that most people don't go to therapists or psychiatrists. Often discourse with a psychic is the first opportunity a caller has had to broach subjects they've avoided or denied. I

also think that deep down most people don't really believe in astrology or tarot the same way they "believe" in psychotherapy. And despite all the mental-health experts they've watched on the *Ricki Lake Show,* a good portion of our society still equates psychotherapy with being crazy or dysfunctional.

Most studies tracing the effect of the sex of a counselor upon clients are inconclusive or are filled with flaws, but a few trends emerge that seem relevant to sex-preferences for psychics:

1. Female psychotherapists report less discomfort during sessions than males.
2. Female therapists seem more sympathetic to the problems of their patients; male therapists were notably harsher in their patient assessments.
3. Female patients report higher levels of satisfaction with women therapists (there is some evidence, certainly not conclusive, that female therapists might be more successful in treating women).

If female psychotherapists are more empathetic to women, perhaps so are female psychics. Why wouldn't women want to consult with a reader who is empathetic and nurturing rather than judgmental or combative?

THE SOCIOLOGY OF PSYCHICS

Historically, the professional psychic has been a female. Sylvia Browne notes that women were the original shamans. In indigenous cultures, it was usually the women who were the healers, the mystics, and the prophets. Even in modern times, it hasn't always been easy for women, especially single women without a formal education, to find work that would provide them with economic independence. James Randi notes that in the nineteenth century and earlier, fortune-telling and psychic work was

> a credible way for a woman to make a living, which is why mostly women became psychics—men would simply get a real job, because they could.

W. Rory Coker feels that these sociological reasons are paramount in explaining the sex differential:

> Historically, this was a career path open to women in the lowest social classes, as well as the low-middle and mid-middle class. Men had many paths to choose from; women did not. The fact that the first mediums were the Fox sisters [who became media sensations in the late nineteenth century; they were eventually exposed as fakes] set the standard. Weren't women more "susceptible" and "weak-willed" and thus easily possessed by spirits? Most female spirit mediums indeed had male spirit guides, as if female spirits were not forceful enough to possess even a living woman.
>
> A number of psychics and spirit mediums I talked to started out as habitual attendees at séances, habitual consulters of psychics. And then they had a kind of epiphany: "I can do this myself, which must mean I have some talent. Since I am not intelligent or educated, the fact that I can give people advice must mean that I am communicating with (usually male) spirits without even realizing it." I have heard this from about half of the fortune-tellers, mediums, and psychics I have had the opportunity to talk to about their background.

Psychic work offers obvious advantages to women, particularly mothers. The work can be done part-time from out of her residence. The hours are flexible, making it convenient for those with childcare or homecare responsibilities.

Most of the female psychics we talked to mentioned that because psychics have traditionally been mostly women, men have been reluctant to enter the field, probably for the same reason that men don't become nurses or elementary school teachers in greater numbers. Even if a man is endowed with talent, he is less likely to enter a woman's "turf." Sylvia Browne hastened to add, however, that although women dominate the psychic field quantitatively, many of the most famous psychics, such as Edgar Cayce, Douglas Johnson, Arthur Ford, and James Van Praagh, are men.

So far, we've been lumping together "intuitives" such as Sylvia Browne along with fortune-tellers, even though there are few similarities in the way they operate. In most urban areas, the field of fortune-telling is dominated by female Gypsies (more properly called "Rom," or "Roma," the word for man/men in the Romani language.) Sergeant Sean McCafferty, of the New York Police Department's Special Fraud Squad, has spent years trying to fight fortune-telling scams. Sergeant McCafferty told *Imponderables* that he has never encountered a male fortune-teller.

Fortune-telling itself is not a crime, but many storefront operations make money by scaring clients with tales of curses that can be cured only by offerings of money or the purchase of special candles. Once a fortune-teller claims to be able to heal a sitter, a crime has been committed. It is not unheard of for sitters to pay from five to thirty thousand dollars to rid themselves of curses. The cynicism with which the Rom women ply their trade is exemplified by one salient fact: Roma rarely go to other Roma to have *their* fortune told.

Not all Roma are engaged in crime, of course, but the pattern for children of fortune-tellers is hard to break out of—girls are usually pulled out of school in the fourth or fifth grade at the latest, and learn how to ply their "trade" by eavesdropping on the palm-reading or fortune-telling sessions of their mothers or grandmothers. As Bob Geis, another specialist in "transient crime," who left the NYPD to become a private investigator, told *Imponderables,*

> In the Gypsy culture, females usually earn the money and the men manage the women. Children are not sent to school but they get a great education. Little girls sit behind the curtains and hear their mothers do the readings, thus learning every trick in the book.

In the Rom culture, it is unmanly to be a fortune-teller, and this belief has feminized the profession in the greater public's eye.

Perhaps women believe more strongly in psychic phenomena. And maybe female practitioners have more talent at "reading" others, or at least more skills in conveying empathy. But our money is on the sociological explanation. Frederick Woodruff sees the obstacles to men joining the ranks of the psychics, but is optimistic about the future:

> Imagine a child telling her father, "Dad, I want to be a numerologist when I grow up," and the dad saying, "Sure! Let me take the money I set aside for your college education and find you a good occultist!
>
> …But this is starting to change. The oppression of the feminine realm is beginning to lift, and these intuitive, reflective, and imaginative qualities are beginning to reassert themselves into everyday life.

The skeptic would look at the same set of facts and observe that as feminine values are more highly regarded, and women are given more opportunities in the workplace, women will have less need for the paranormal to help them cope with everyday life. While this transition is taking place, there will still be more female sitters and readers.

Submitted by Kathryn Rutherford of Grissom AFB, Indiana. Thanks also to Dennis Estes of Cascade, Maryland.

Who Was Monterey Jack, and Why Is a Cheese Named After Him?

You don't know Jack? Neither did reader Phil Hubbard, who admitted that he first submitted this Imponderable as a joke, "assuming that this was still another use of the word "jack" that does not refer to an actual person."

But then Phil went to Merriam-Webster's Web site and found that the dictionary ascribed the etymology to an "alteration of David Jacks, nineteenth-century California landowner." How did a landowner in Monterey become associated with California's most popular native cheese?

David Jacks, a Scottish immigrant, moved to California during the gold rush in the mid-nineteenth century, where he amassed wealth by selling dry goods to miners. Jacks gobbled up land and quickly became one of the largest landowners in Monterey County.

Jacks was known to lend money to insolvent borrowers, using their land as collateral: He was not shy about foreclosing.

Jacks was not a cheese maker, but he did acquire approximately fifteen dairies as part of his real estate portfolio. Although there is some controversy about the etymology of Monterey Jack, the California Dairy Research Foundation's explanation is the most widely dispersed:

> As the story goes, sometime in 1882 David Jacks began shipping from his dairies a cheese branded with his last name and the city of origin, Monterey, to San Francisco and other western markets. Eventually the "s" was dropped and people began asking for "Monterey Jack." While there are alternative explanations for the cheese's origins—such as that the cheese was first made using a "jack" (or press)—David Jacks is the one most often credited for its distinctive name.

Although Jacks might have given the cheese its popular name in the United States, chances are Monterey Jack is a descendant of cheeses that were first brought to California by Spanish missionaries a century before. The missionaries called their cheese *queso del pais* or "country cheese." In the 1850s, Dona Juna Cota de Boronda, a housewife with a disabled husband and fifteen children, started selling her queso del pais door-to-door in Monterey. Those who credit Boronda with being queen of Jack cheese claim that Jacks usurped a name already in circulation, because of the press used to make the cheese. Judge Paul Bernal, San Jose, California's official historian, is clearly in this camp:

> Some consumers looking for Boronda's cheese would ask for the "jack" cheese (cheese made with a press or jack). Some

would ask for Monterey Cheese. Capitalizing on the confusion of terms and producers, David Jacks cleverly renamed his brand "Monterey Jack Cheese" so all buyers would gravitate toward his cheese. Of course, Boronda was wiped out and Jacks became wealthy, enabling him to build the Jose Theatre [built in 1904 and still standing], among other enterprises.

There are other claimants to the throne of originators of the cheese, all of whom argue that the jack in question refers to the press used to make the cheese. About the only group not trying to claim credit are the Franciscan monks who created the cheese in the first place.

Thanks to Phil Hubbard of Williamsburg, Kentucky.

Why Is the Moon Sometimes Visible During the Day?

This Imponderable would be *so* easy to answer if the sun, moon, and Earth would get together and agree on a uniform schedule. But they refuse to do so, keeping astronomers and astrologers in business, and making it hard for us to provide a simple answer.

Here's the simple answer, anyway. The moon does not shine by its own light. When we see the moon, it's only because we are seeing the reflection of sunlight bouncing off its surface. You can see the moon in the daytime when the sun and the moon are located in the same direction in the sky. As the moon proceeds on its (approximately) twenty-nine-day orbit around the Earth, at times it's on the opposite side of the Earth from the sun.

Although we may remember this only when we've been indulging in too many recreational substances, the Earth is also spinning on its axis once every twenty-four hours, a much shorter time than it takes the moon to revolve around us or for the Earth and moon together to orbit around the sun (a year). Though we perceive the moon as rising and setting as it "moves" across the sky, it's really the Earth rotating on its axis ("underneath" the moon) that causes this effect. In one day, the moon doesn't move much relative to the sun or to the Earth, even though during these twenty-four hours, we see a complete cycle of day and night because of our planet's spinning.

Viewing of the moon is also contingent on the state of the Earth's atmosphere. The stars are "out" during the day, but we can't see them because the scattered light from the sun is bright enough to drown out the relatively dim light from the stars. But the moon is the second brightest object in the sky, next to the sun, so even though it appears pale, we can usually see it during the day if it is close in direction to the sun. But on days with excessive glare or cloudiness, the moon may not be visible, especially just before and after a new moon.

Even though the moon and sun often appear to be close together, the sun is always about 400 times farther away from Earth than the moon. We can see the moon during the daytime when the sun and moon are relatively close in direction, but not too close! When they are aligned too closely, we can't see the moon because the sun is directly behind it and can't light up the side of the moon facing us. When they are in opposite directions, in the daytime, the sun is overhead but the moon is on the opposite side of the Earth.

When the moon is overhead, you do see it, but it is night because the sun is on the other side of the Earth. It's when the moon and sun are at right angles, or close to it, that you can best see the moon during the day—the sun, moon, and Earth form a big triangle, and the sun is "in front" of the moon to light up the side of the moon that is visible to us, and it's daytime because the sun is up in the sky above us.

Still confused? Maybe this analogy from Tim Kallman, an astrophysicist at the Laboratory for High Energy Astrophysics at NASA, will help:

> It might be useful to think of the sun as a large light bulb, and the moon as a large mirror. There are situations where we can't see the light bulb, but we can see the light from the bulb reflected in the mirror. This is the situation when the moon is out at night. We can't see the sun directly because the Earth is blocking our view of it, but we can see its light reflected from the moon. However, there are also situations where we see both the light bulb and the mirror, and this is what is happening when we see the moon during the day.

Submitted by Glen Kassas of Concord, New Hampshire. Thanks also to Caroline and Cathy Yeh, of parts unknown; Margaret Paul Vitale of Palermo, Italy; and Terry Keys, Jr. of Friendswood, Texas.

DAVID FELDMAN

How Do They Get the Paper Tag into Hershey Kisses? And Why Are They Called Kisses?

The origins of Hershey Kisses lay more in technology than romance. Before Kisses, Hershey sold a molded candy called Sweetheart. It was cone-shaped and featured a kiss imprint on its base. No one knows for sure if Sweetheart inspired the Hershey Kiss. According to Pamela C. Whitenack, director of the Hershey Community Archives, and the source for most of our information in this chapter, the word "kiss" was already a common confectionary term for small candies when Hershey first marketed its Kiss in 1907.

The key attribute of the Kiss was its distinctive shape, and the difficulty in its production was developing machinery that could extrude chocolate at the proper temperature, and then cool it quickly so that the whirl on top remained intact. Hershey wasn't able to figure out how to wrap Kisses by machine at first—every Kiss was individually wrapped by hand, with a small tissue (containing the Hershey trademark) surrounding the chocolate and housed inside the foil exterior. In *The Emperors of Chocolate: Inside the Secret World of Hershey and Mars*, Jöel Glenn Brenner points out the problems with this method:

> To wrap a single Kiss was a delicate process: the tissues inserted in each one had a tendency to blow away, and were difficult to handle. A proper wrap required picking up the tissue, laying it on a foil, placing the Kiss on top and twisting the whole package together. But this process took too much time. Some workers were known to pick up a Kiss, lick the bottom, dab it on a pile of tissues, then deposit that on the foil and twist.

Another weakness of the early Kisses was that there was no clear indication of the Hershey name on the exterior. Kisses were first sold in bulk at approximately thirty cents per pound. Hershey needed a way to make its confections distinctive and identified with its brand, and the solution came in the form of the little tag, which Hershey calls a "plume." The plume was made possible by the creation of a suitable wrapping machine in 1921.

Although the single-channel wrapping machine has since given way to a complex wonder that can wrap up to 1,300 Kisses in an hour, even the current machine essentially duplicates the process of the hand wrappers. The foil and the plume material are brought to the wrapping area in continuous rolls, and then threaded separately through the wrapping machine so that the plumes are placed on top of the foil. The two materials are then precision-cut to exact specifications, so that the plume pokes its head out of the foil. Naked pieces of chocolate are centered on the foil-plume combination and the wrappers are twisted before exiting the machine. Then the finished individual Kisses are sent to another station for inspection, weighing (there are ninety-five Kisses to a pound) and bagging. When multiple color foils are used, such as for holiday Kisses, the additional foils are blended together at this stage.

Where did the name "Kiss" come from? No one seems to know. Pam Whitenack told *Imponderables* that although "kiss" was used to describe bite-sized candies in the nineteenth century, it didn't stop after the introduction of the Hershey Kiss, either:

> I have a page from the *Confectioner's Journal* (a trade publication from the turn of the twentieth century) that shows more than two dozen different kinds of confectionary kisses. Jolly Rancher Bites were marketed as "Kisses" prior to the company's acquisition by Hershey Foods Corporation. The product name was changed to avoid confusion with Hershey's older and more recognizable product.

It was not until 1923 that Hershey obtained a registered trademark for Kisses, and that wasn't for the name alone, but for the "basic shape, size, and configuration of Kisses, with its foil wrap."

Is it a coincidence, corporate espionage, or cosmic fate that in the same year that Hershey Kisses were introduced in the United States, 1907, Perugina launched its line of small chocolates, Baci, in Italy? The Italian chocolate also sports a similar but not identical swirl on top. Baci offers no plume, but includes a love note (in four languages) inside every chocolate. Oh, one other thing: *Baci* just happens to mean kisses in Italian.

Submitted by Anthony Cusumano of Ashburn, Virginia.

How Does the Vending Machine Know When It's Sunday and That the Newspaper Is More Expensive?

Never underestimate the intelligence of a newspaper vending machine. It knows darn well when you try to pay the price for a daily newspaper on Sunday. You can try to get away with inserting fifty cents when you owe two dollars, but our money is on the vending machine shutting you out.

When you insert coins into the vending machine, the quarters don't just drop into an empty vessel, but into a mechanical or electronic receptacle ("mech"). Even the more primitive machines can account for at least two different price structures—most commonly, to account for a more expensive Sunday paper. As Bob Camara, sales representative for K-Jack Engineering Company put it: "The coin mechanism is changed over to a Sunday setting."

The people who service the racks reset all the mechanisms to a Sunday setting to assure that bargain hunters are frustrated. Pamela Davis of Rak Systems, Inc., explains:

> There are mechanisms that have to be set manually each time the price changes. There are also electronic mechs that can be changed with the click of a button. Some of the electronic mechs can also collect data . . . what time and day papers are bought, what coins are used, etc., to help the newspapers know what their best markets are.

> A collection of Rak's mechanisms are on display at http://www.raksystems.com/mechs.htm. There you'll see that the preoccupations of Rak's clients are few but persistent—multiple pricing options, ability to collect coins other than quarters, and the most nagging issue since the advent of the vending machine: rejection of slugs.

Submitted by Jim Barton of Phoenix, Arizona.

DAVID FELDMAN

What Happens to Olives After the Oil Is Squeezed Out?

Lucy Ricardo might have been able to crush grapes with her feet to make wine on *I Love Lucy,* but extracting olive oil from olives is a little trickier than juicing a grape or an orange. Olives don't contain as much moisture content as grapes, and possess nasty, hard pits.

The first important process in making olive oil, after separating the olives from dirt, leaves, and other contaminants, is pushing them through a mill or grinder, which turns them into a fine paste. Usually, the olive pits are left in before processing; the pits don't have much effect on flavor, and contribute little to the volume of oil—but "destoning" olives adds another step, and expense, to the process.

After the olives are mashed into paste form, the paste is mixed ("malaxation") for about one half hour, which allows larger bubbles of oil to coalesce. The next stage is crucial—extracting the oil from the water in the paste, usually accomplished by one of two types of machine: a press or a centrifuge.

Although the valuable oil has been extracted, and the olives have lost all of their original texture, olive oil producers don't toss the waste product, which is called "pomace." Paul Vossen, a University of California farm advisor, told *Imponderables* about the fate of the pomace:

> The pomace can be used for compost, or if the pit fragments are removed and it is somewhat dried, it can be fed to livestock. In Europe, North Africa, and the Middle East, the pomace is placed into solvent tanks and the remaining small amount of oil is removed; the solvent-extracted oil is refined and sold as pomace oil. The spent pomace is usually burned to generate heat and dry the pomace before it is solvent-extracted.

The pomace olive oil is controversial in the trade. As we discussed in *When Do Fish Sleep?*, virgin and extra virgin olive oil are prized not only for their low acidity and fine taste, but for the lack of processing used to extract them. Purveyors of fine olive oils tend to look down on solvent-extracted pomace oil, especially because pomace can legally be labeled "olive oil." Betty Pustarfi, owner of Strictly Olive Oil in Pebble Beach, California, refers to pomace olive oil as "industry denigrated oil" because it is used to:

> mix itself with the *real stuff* so it can be sold to consumers as premium, or most frequently, sold or used to be blended with the *real stuff* for use in the food service or production

industry. Pomace olive oil is a lubricant, not a condiment, though it has most of the health values of the *real stuff* and is an accepted carrier for the *real stuff* as long as it is so labeled.

Submitted by Rene Triliad, via the Internet.

For more information about the making of olive oil, visit the Olive Oil Source Web site at: http://www.oliveoilsource.com/making_olive_oil.htm.

Do Ostriches Swim?

We all know that ostriches don't really bury their heads in the sand. But will they dunk their bodies into the water? We vaguely remembered Charles Darwin writing about swimming ostriches, and we found the passage that we remembered in *The Voyage of the Beagle*. The scientist was traveling through northern Patagonia and discussed the South American ostriches he observed, but the description was a little more elusive than remembered:

> at Bahia Bianca I have repeatedly seen three or four come
> down at low water to the extensive mud-banks which are then
> dry, for the sake, as the Gauchos say, of feeding on small
> fish. . . . It is not generally known that ostriches take to the
> water. Mr. King informs me that at the Bay of San Blas, and

at Port Valdes in Patagonia, he saw these birds swimming several times from island to island. They ran into the water both when driven down a point, and likewise of their own accord when not frightened: the distance crossed was about two hundred yards. When swimming, very little of their bodies appear above water, their necks are extended a little forward, and their progress is slow.

So far, Darwin's account is second hand, it would seem. But eventually, he recounts:

On two occasions I saw some ostriches swimming across the Santa Cruz river, where its course was about four hundred yards wide, and the stream rapid.

Case closed? When we conducted some further research, we were taken aback to read that ostriches are native only to the savannas and deserts of central and southern Africa. The "South American ostriches" that Darwin observed were not ostriches at all, but close relatives, the rhea (they belong to the same order as ostriches, Struthioniformes, but a different species). Rheas look similar to ostriches but with more water sources available to them than ostriches in the desert, might they have abilities in the water that ostriches don't have?

The first person to respond to our query was Prof. Gerhard H. Verdoorn, director of BirdLife South Africa, who wrote:

I have no evidence from any literature that ostriches can swim! So the answer from my side is no! I guess that they will have to swim if their area becomes inundated—otherwise, no records of them swimming.

Next in our queue was an e-mail from Steve Warrington, of Ostrich.com:

> Yes, ostriches can swim—in their natural desert surroundings they swim a lot—usually to cool and clean themselves.

A South African ostrich farmer, Tania Lategan from Cango Ostrich Farm outside of Oudtshoorn, South Africa, responded matter of factly, "Yes, an ostrich does swim to cool its body off." We asked her if her ostriches swim on a regular basis, and she said: "Yes, if they do have access to water and it is hot, it will occur every day."

But then another source, Pierre duP Fourie, proprietor of the Baron's Palace Hotel, just miles from Cango Farm, reports:

> When forced into a flooded river, they float in the water to safety, but do not go into water to cool down or swim.

But veterinarian Carole Price, president of the American Ostrich Association, counters with:

> Yes, ostrich swim. In fact, they *love* water.

So here we are answering this Imponderable with a qualified yes: Ostriches certainly *can* swim, but perhaps reluctantly so. We were able to find one image of an ostrich swimming at http://www.marijuana.org/pictures6-14-99/swimingostrich.JPG, but we're not sure what role marijuana played in the proceedings!

Submitted by Galen Musbach of Greeneville, Tennessee.

DAVID FELDMAN

When You Switch Chairs with Someone, Why Does the Seat Sometimes Feel Warm?

This is a rare example of an Imponderable posed in front of us at the moment it was born. A pal, Chris McCann, was playing West in a duplicate bridge tournament. In duplicate bridge, the North-South players remain seated at the same table, while East-West players move from table to table after playing a few hands. When Chris came to our table, he noted that the wooden chair he sat in felt very warm.

We have had the same experience, so for years we've done some empirical research on this Imponderable. We also consulted with two science types, both of whom hold advanced degrees from accredited universities, and because they have a reputation to protect, didn't rush to have us quote them by name. They confirmed our impressions that this is a matter of simple, thermal conductivity: When two surfaces press against each other, there is an exchange of heat between them. But what determines the degree of heat transfer?

1. Time and Temperature

Although the core temperature in our body is approximately 98.6 degrees Fahrenheit, the temperature of our skin is lower, usually a little above 90 degrees. Under normal circumstances, we're sitting in rooms that are much cooler. So if a chair has been previously occupied, the former occupant, let's call him "Ex," is going to warm a chair to a higher temperature than the ambient air in the room.

When we are first seated in a restaurant, though, we are unlikely to perceive our chair as warm, because the table has been unoccupied long enough for the chair to cool off to the room temperature. At the bridge tournament, the new occupant, let's call him "Chris," might be in Ex's seat within seconds of Ex leaving, allowing plenty of time to bask in the warmth left behind.

2. Length

The longer Chris has been standing up, exposed to the 70-degree room, the more likely he's going to feel that the chair that Ex is vacating will be warm. The longer Ex has sat in the chair, the warmer that chair will be, as more of Ex's heat will be transferred to the chair. But the longer Chris sits in the chair, the less he is likely to continue *perceiving* the chair as warm, as the contrast in temperature peaks upon first contact.

3. Width

There is no getting around it. There is a direct correlation between the mass of the buttocks region and how much heat Ex is able

DAVID FELDMAN

to transfer from his booty to the seat. The wider the rear, the tighter the seal between the skin (with its higher temperature) and the surface of the chair is accomplished, and the more heat is transferred. If Ex filled out the seat before him, even a slimmer Chris is sure to hit a "hot spot." There are plenty of bridge players with ample rears, and in our experience they tend to leave the gift of warm seats when they move on to the next table.

4. Surface

What kind of clothing are Ex and Chris wearing? The thinner the clothing, the more easily heat can be transferred to or felt from the chair. Probably even more important is the composition of the chair. Metal is a great conductor. Ever sit on an aluminum chair in shorts in an air-conditioned room in the summer? The chair will feel much, much cooler than a chair with fabric on the seat and backrest, because the metal chair leeches heat out of our bodies quickly. Textiles are poor conductors of heat, as they leave plenty of room for air to be trapped inside, and stationary air doesn't promote heat transfer. This is why bulky sweaters are effective in the winter—the trapped air in the material helps to insulate you from the cold. Wooden chairs are in between fabric and metal, but closer to fabric in conductivity.

5. Psychology

No doubt our mind plays tricks on us. If you expect a metal chair to be cold, and the Ex who preceded you, with ample rump covering the whole seat, has sat in it for a long time, the chair might feel surprisingly warm, even though in actuality its surface temperature is lower than the fabric chair alongside it. And perhaps, if you expect a warm seat, it will feel warmer than it really is.

But if you want to insure a warm seat, watch for the big guy with the wide booty and the thin clothes. As soon as he leaves, hasten thee to that chair.

Submitted by Christopher McCann of Brooklyn, New York, who exhibits low butt mass.

What Accounts for the Different Shapes of Cheeses? Why Is Cheddar Rectangular While Brie and Provolone Are Round?

E ver since we inquired into the origins of why there are ten hot dogs in a package and eight hot-dog buns (even we don't drop enough franks off the grill to justify the short-fall), we've been wary of finding any logic in the world of food packaging. We were a little surprised at how many people within the cheese and dairy industry couldn't answer this Imponderable. But luckily we found two experts who could: David Brown, senior extension associate at Cornell University's food science department, and Dean Sommer, a cheese technologist at the Wisconsin Center for Dairy Research in Madison, Wisconsin.

Traditionally, most cheeses were made in round form, and this includes most of the cheeses we find now in rectangular form, such as Cheddar and Swiss. The classic Cheddar was made in forty-pound wheels, and in England, most Cheddars still are round. The Swiss often created much bigger wheels, as heavy as 200 to 220 pounds. Sommer points out that even in the United States, you can see vestiges of these traditions in half-round Cheddar and in colby longhorns or Cheddar longhorns. For these popular cheeses, the shape has no influence on its flavor or texture. If you look at recipes for making cheese, the shape of the finished product is usually optional.

So if one shape isn't inherently superior in making better cheese, what accounts for the rectangles we see on grocery shelves? One reason that both our experts mentioned was that rectangular packages were easier to stack on grocery shelves, with less wasted space than round ones.

But more important is that the "conversion" from big blocks of cheese to consumer-sized pieces be accomplished with the least amount of scraps. Sommer wrote us:

> Converters have a lot of waste when they cut a round piece of bulk cheese into rectangular retail pieces of cheese. So the cheese industry converted their Cheddar production for the most part from various round shapes (flats, daisies, longhorns, midgets) over to rectangular forty-pound blocks (the standard shape for the industry) or even rectangular 640-pound blocks to minimize trim losses and to maximize efficiencies in cheese production, conversion, and distribution processes.

Brown adds that sometimes retail accounts, such as gourmet stores, ask for five- to ten-pound bricks (to sell as is or from which to cut smaller pieces), and these are also more easily created from rectangular bulk cheese.

But the round cheese is far from extinct. Because round cheeses are associated with traditional methods, most artisanal cheese is made in this shape, even when there is no technical reason to do so. But some cheeses are more easily made round, as Sommer explains:

> Smear ripened cheeses like Limburger and brick cheese can only be made in relatively small sizes due to the need for the enzymes that form on the surface of the cheese to migrate to the very center of the cheese over time. If the cheese is too large, then the enzymes cannot reach the center in a reasonable curing time. Similarly, in cheeses such as Camembert and Brie, the mold growing on the exterior of the cheese produces flavors and enzymes that need to migrate into the interior of the cheese. With blue cheese, the round shape is optimal for brining and salt absorption and if the cheese isn't in a relatively round, small wheel the inward pressures would be too great, collapsing the open structure and not allowing the blue mold to grow as well (because the mold needs air pockets inside the cheese to grow).

Many hard cheeses, especially ones that are brined or smoked, work best when made in a wheel or cylinder, where the flavoring on the outside, whether salt or brine, can evenly penetrate the interior. Provolone was traditionally made in the shape of salami. Authentic provolone is smoked, and David Brown speculates that the 200-pound provolone might have hung next to salamis and other cured meats in smokehouses in Italy.

Mozzarella was traditionally created in balls, probably to

promote evenness in the brining process, and you still see them in Italian markets and gourmet stores, but because it has become a mass-market item, you can find rectangular specimens in stores, along with the shredded mozzarella for lazy pizza makers.

Submitted by Julie Erskine of Columbus, Ohio. Thanks also to Zoe Klugman of Guilford, Connecticut.

What Does the "D" in D-Day Stand For?

On June 6, 1944, 156,000 Allied soldiers headed to the shores of France (most famously, in Normandy), as part of Operation Overlord, the code name for the entire Allied invasion of northwest Europe. Not all of the soldiers landed on the beaches on June 6, but that day became known as D-Day, the beginning of the pivotal Battle of Normandy.

In an unscientific sampling of friends and acquaintances, we received all kinds of guesses about what the D might represent. Some of the guesses included: Decision, Disembarkation, Debarkation, Doomsday, Deliverance, and most commonly, Dunno.

We contacted the D-Day Museum in Portsmouth, England,

and a representative wrote *Imponderables* that the museum's own Web site's explanation was as good as any:

> When a military operation is being planned, its actual date and time is not always known exactly. The term "D-Day" was therefore used to mean the date on which operations would begin, whenever that was to be. The day before D-Day was known as "D − 1," while the day after D-Day was "D + 1," and so on. This meant that if the projected date of an operation changed, all the dates in the plan did not also need to be changed. This actually happened in the case of the Normandy landings. D-Day in Normandy was originally intended to be on 5 June 1944, but at the last minute bad weather delayed it until the following day. The armed forces also used the expression "H-Hour" for the time during the day at which operations were to begin. . . .

Both the U.S. and British military have the same designations for "D" and "H" in military planning. We haven't been able to find its first use in England, but in the United States it dates back at least to World War I. According to the U.S. Army Center of Military History,

> The earliest use of these terms by the U.S. Army that the Center of Military History has been able to find was during World War I. In Field Order Number 9, First Army, American Expeditionary Forces, dated September 7, 1918: "The First Army will attack at H hour on D day with the object of forcing the evacuation of the St. Mihiel Salient.

Submitted by Lance Tock of Brooklyn Park, Minnesota.

DAVID FELDMAN

Why Is an Elephant's "Nose" Called a Trunk?

Although an elephant's trunk seems to be wide as the back of a Lincoln Continental, the big lugs' proboscises are named after their resemblance to a tree's trunk. As veterinarian Myron Hinrichs wrote *Imponderables*:

The elephant's trunk looks just like the trunk of a tree, thick and broad at the base and more slender at the tip.

The first recorded use of the word "trunk" to describe an elephant's nose was in 1565, in a translation by Richard Eden, an English translator of many books on travel, geography, and navigation. According to the *Oxford Dictionary of English Etymology*, "trunk" was first used to describe the main stem of a tree only a century before.

Submitted by Michael Green of New York, New York. Thanks also to Jennifer Erin Hester of Albany, Georgia.

DAVID FELDMAN

Why Are American Football Fields 53 ⅓ Yards Wide?

I n *Do Penguins Have Knees?*, we answered a similar Imponderable about the weird distance between the pitcher's mound and home plate (sixty and one-half feet), and concluded that most likely the culprit was a misread of architectural drawings that called for a nice, even sixty feet. Although no one knows for sure, including the National Football League and the Pro Football Hall of Fame, our guess is that some faulty arithmetic might be to blame for the weird asymmetry of the football field.

Bob Carroll, executive director of the Pro Football Research Association, is our go-to guy about football history, and he filled us in on the ever-shifting dimensions of football fields. In the early days of football, the field was 420 feet by 210 feet—a tidy 2:1 ratio. But in 1881, the length was reduced to 330 feet or 110 yards. All of this length was used for the game itself; end zones did not exist because the forward pass had not yet been introduced. But in 1881, the width was not reduced to exactly half the length; 160 rather than 165 feet was chosen as the width, possibly because it was an even number, possibly because someone hadn't done a very good job comprehending their pre-algebra lessons. The playing area was later reduced to 100 yards, with two ten-yard end zones, but the damage had been done: The 2:1 ratio of length to width was a thing of the past.

Submitted by John Martin of Sebastopol, California.

Why Do Books, Legal and Financial Documents, Manuals, Pamphlets, Musical Scores, and Standardized Tests Sometimes Have Pages That Say: "This Page Intentionally Left Blank"?

Almost ten years ago, we received a thick envelope from Bill C. Davis of Portland, Oregon. He posed this Imponderable and mentioned that he stumbled upon a book that was a "gold mine of pages left intentionally blank." The book, an instruction manual for the Microcom QX/2400t Error-Correcting Modem, might not have been a bestseller, but it was a perfect illustration of Bill's point. He noted that the blank pages were numbered and each blared "This Page Intentionally Left Blank," with each word capitalized.

For proof, he included a sample of each of the blank pages:

> I copied these pages at PIP Printing in Portland. The lady
> at the counter was looking at me funny. I would guess that
> not that many people come in to copy intentionally left blank
> pages! Her name is Margaret and I asked her if she remem-
> bered seeing intentionally left blank pages before. She said
> that printers will print pages in this manner for their pur-
> poses and that leaving this on the page may be a mistake.
> I think what she was saying was that it may not be inten-
> tional to print "This Page Intentionally Left Blank" on the
> page!

For Bill, Margaret, and the other readers who have pondered this
mystery, we're here to help.

As we wrote in our first volume of *Imponderables,* most books
have a few empty pages at the front or the back. We in the publishing biz are cool enough to let them bask in their nakedness. But let
the lawyers, financial titans, and technical writers and editors get involved, and weirdness ensues. Englishman Guy Chapman, who has
seen a technical manual or two in his days in the computer industry,
wrote a perfect preamble to this Imponderable:

> "This Page Intentionally Left Blank" could be one of the
> oddest sentences in the English language. Found in instruc-
> tion manuals around the world, it indicates that the page on
> which it appears has been purposely left empty of words or
> pictures. But once this phrase has been printed on the page,
> the page is no longer blank; in fact, it is *intentionally* not
> blank. Therefore, this statement is only correct when it has *not*

yet been made. Once it is written down, it is instantly wrong. By virtue of self-reference, the phrase is denying its own existence and contradicts itself. The only known phrase that is more confusing is "This is a lie."

There are slight variations to the wording. Every writing teacher tells students to make their writing punchier by using more verbs and fewer adjectives and adverbs. Some choose to transform the dowdy "This Page Intentionally Left Blank" to the downright adventurous and sinewy, "This Page Is Intentionally Left Blank." Chapman found one manual writer who was so chagrined at the thought of the TPILB paradox that the bereft page was marked with "The page on which this statement has been printed has been intentionally left such that this statement is the only statement printed on it." That writer, evidently, was paid by the word.

Reader Harvey Kleinman opened up the prospectus for a mutual fund, Vanguard Windsor II, and realized that the last page (not the cover or inside back page) contained these magic five words. He noted that just as books are not printed on individual sheets but on larger collections of eight, sixteen, or thirty-two pages, the prospectus was a folded four-sided piece of paper. Because the prospectus is a legal document, which must be issued to investors, it wasn't the right venue to casually insert an advertisement for other funds, a random illustration, or a photo of Catherine Zeta-Jones. As Chapman points out, in some instruction booklets, you will find "Notes" as a header on otherwise empty pages, which makes it look like the producer cares enough about you to provide you with a writing area, when in fact it is to hide the embarrassment of a naked page.

TPILBs have been found in a myriad of documents. Let's look

at some of them, and find out why they appear:

Legal Documents: Let's say you are an associate at a big law firm, assembling a lengthy brief (oxymoron intended) full of documentation to support your case. You paginate these hundreds of documents and then at two in the morning, your managing partner calls and barks: "You know that Perlman affidavit? Yank it!" You follow orders, but what do you do with what are now incorrectly numbered pages? Before computerization, repaginating was a nightmare—now it is easier. But according to the lawyers we consulted, it is wiser to insert a TPILB than to try to repaginate, and not just because of the time wasted or the drudgery of reworking the numbers. According to Ed Swanson, an attorney in Los Angeles specializing in corporate and securities law, a lawyer has to worry about the accuracy of every cross-reference and page citation throughout the document.

Swanson mentions that in documents that require signatures, lawyers usually want all the signatures to be on the same page, and separate from other material. Or they might want to highlight a particular heading or caption. Depending upon the length of the material preceding it, the important material might best be accentuated by having an empty page preceding it—time to drag out the trusty TPILB—or sometimes just an "Intentionally Left Blank," if most but not all of a page is empty. Why bother with the warning message? Lawyers are paid to contemplate every awful, even if unlikely, eventuality, and a nefarious type could insert unwanted material into the empty space. The "ILB" makes it clear, Swanson observes, that "Nothing funny is going on." TPILBs occasionally appear in classified documents, especially military publications. Obviously, this is a sensitive area where the consequences of "nothing funny is going on" are anything but funny.

Sometimes, a TPILB is the result of sloth rather than meticulousness. We cajoled a lawyer into admitting that he has created a few TPILBs in his time, and he admits that sometimes they are put in documents out of "sheer laziness." Paul Dahlman, an attorney in New York City, e-mailed us:

> The scenario runs something like this. When negotiating a
> commercial lease, the landlord's lawyer sends over a proposed
> lease with 30 to 50 pages of riders to the standard form lease.
> Because it is cut and pasted (often physically pasted) and is an
> attempt to customize the lease, the lawyer simply cut out
> pages that didn't fit, and hit print.

Financial Documents: Most of the factors stated above apply here, which makes sense, as lawyers have their way with financial documents, too. Jack Suber, a lawyer and general manager of American Financial Printers, adds that one difference in financial documents is a greater incidence of right-hand TPILBs, which we've noticed in annual reports and prospectuses from financial institutions:

> Intentional blanks are almost exclusively used on the left-
> hand page but do occasionally show up on the right if a series
> of financial tables is laid out in such a way that two pages
> must be across from each other to create an extra-wide table.
> In that case, an "intentional blank" is sometimes inserted be-
> fore those pages as a sort of stutter-step so that tables that
> would have backed up to each other can face each other.

Customized Booklets and Manuals: When you look at an electronics manual these days, often it will cover many different variations of the same model. It is easy to mistakenly read the right advice for the wrong product. That's one major reason why, in the era of desktop publishing, many corporate customers demand custom manuals for the products and services they buy, ones that cover the exact configuration they have purchased. It's a snap for the publisher to patch together modular, preprinted chapters, and even to provide custom pagination. But some of these chapters will end on a right-hand page, and the custom is for all new chapters to start on the right. Solution? TPILB.

SATs and Other Standardized Tests: TPILB or "NTCOTP" (No Test Content On This Page) lets test takers know that they have reached the end of one section of these timed, standardized tests. The wrath of the College Board will be unleashed on the test taker who "mistakenly" flips beyond to the next section of the test before so instructed.

Musical Scores: Imagine pounding on the piano, finding the time to flip the page in the midst of the performance, and seeing a blank page staring at you. Sheet music publishers try to pattern scores so that the fewest page turns are necessary, including defying publishing tradition by starting compositions on the left-hand page. TPILB lets the musician know that all is well.

Jack Suber believes that the majority of TPILBs occur because of our tradition of starting new chapters, and page one, on the right-hand side. Doing otherwise is so alien to customary practice that Suber says the finished product is called a "Chinese folio." As we discussed in *Why Don't Cats Like To Swim?*, this custom started before there were covers on books. If there was no binding, then there *was* no left-hand page at the start of a book. The technological reason for the right-hand first page no longer exists, but it remains as a curious vestige of the bygone days of printing.

But what if there is no left-hand page in a document today? TPILB has infiltrated the World Wide Web. The "This Page Intentionally Left Blank" Project fears that in the computer age, TPILB might be an endangered species, so it has issued this manifesto:

> In former times printed manuals had some blank pages, usually with the remark "this page intentionally left blank." In most cases there had been technical reasons for that. Today almost all blank pages disappeared and if some still exist here and there, they present flatterly [sic] comments like "for your notes" instead of the real truth: This page intentionally left blank!
>
> Nowadays the "This Page Intentionally Left Blank" Project (TPILB Project) tries to introduce these blank pages to the Web again. One reason is to keep alive the remembrance of these famous historical blank pages. But it is the primary reason to offer Internet wanderers a place of quietness and simplicity on the overcrowded World Wide Web—*a blank page for relaxing the restless mind.*

We're proud to do our part to spread the five magic words, at http://www.imponderables.com/tpilb.php.

Submitted by Bill C. Davis of Portland, Oregon. Thanks also to Chris Curtis of Denver, Colorado; Harvey Kleinman of New York, New York; and Jonathan Ah Kit of Lower Hutt, New Zealand.

In Bowling, Why Is a Strike on the "Wrong" Side of the Headpin Called a "Brooklyn Strike"?

Most bowlers are right-handers, and their tendency is to throw the ball with a natural hook, with the ball moving from right to left. Right-handers with hooks aim the ball just to the right of the headpin (the "one pin") so it simultaneously hits the three-pin, too. If the ball hooks a little too much to the left, and the ball knocks the headpin straight on, a dreaded, impossible-to-convert split is often the fate.

But when the right-hander misses by a greater margin and the ball heads to the left of the one-pin, the bowler often lucks out with a strike, even though the target has been missed by a wider margin. Professional bowlers are sheepish when they've "achieved" a Brooklyn strike; if you want to irritate serious bowlers, have an opponent of theirs win a match by employing one. As one frustrated amateur admitted on an online bowling bulletin board:

> I never give a high-five to a Brooklyn strike. On the times when it's been offered by the bowler, I've simply told them they have to do better than that to get a high-five from me.

How was Brooklyn chosen to designate this errant but lucky strike? Mort Luby, publisher of *Bowlers Journal International,* wrote *Imponderables:*

> Brooklyn was considered the wrong side of town. Thus, strikes resulting from balls striking the "wrong" side of the headpin were so-named.

Who would have enough of an "attitude" to make fun of Brooklyn in this way? Of course, it's the New Yorkers who think they live on the right side of the town: Manhattanites. And if you need further evidence that Luby's theory is correct, keep in mind that although the term "Brooklyn strike" also applies to left-handers who knock down ten pins by hitting the one-three pocket instead of the desired one-two, another term used to describe a lefty Brooklyn strike is a "Jersey strike," traditionally New Yorkers' other favorite location for barbs.

Submitted by Michelle Marsaglia of Salem, Oregon.

Why Are the Number 13 and Friday the Thirteenth Considered Unlucky?

lthough these are two of the most frequently posed mysteries by readers, we've resisted answering them for a couple of reasons. When in doubt, we try not to use mysteries that can be answered only by other books. But since we can't travel back in time, nor channel the long-deceased to answer this Imponderable, we are stuck with written sources.

Most of the books we have consulted leave us frustrated. There are literally scores of books about superstitions, and just about all of them address the fear of 13. Most of them contend that the fear of 13 stems from the Last Supper, where Judas was the thirteenth guest to sit at the table.

The other most common theory is that the superstition predates Christianity, and is based on an ancient Norse legend in which

Baldur, the god of light, is killed by the evil Loki. In a story quite reminiscent of the Last Supper, twelve gods are dining in Valhalla when they are "crashed" by the evil Loki. Baldur is killed soon afterward, because of the plotting of Loki.

Most books about superstitions assume that Friday is particularly reviled because it was the day of the Crucifixion. In other variations, it is the day that Adam ate the apple.

But there are problems with all of these theories and we thought the arguments were too shaky to include in an *Imponderables* book. Then one day, while visiting one of our favorite bookstores—the Tattered Cover in Denver, Colorado—a book with the title of *13* caught our attention (and not just because it happened to sit next to *Do Elephants Jump?*—we would *never, ever* go to a bookstore just to check how a book of ours is selling). Written by Nathaniel Lachenmeyer, *13* is a fascinating cultural history of "the world's most notorious superstition." In the book, Lachenmeyer articulates our misgivings about prior explanations, and through meticulous research, offers informed opinions about the origins of triskaidekaphobia (the fear of 13).

Lachenmeyer swats away most of the conventional wisdom. Yes, there is a Norse legend of Baldur, but there were actually 13 gods, not twelve, when Loki appeared on the scene, so 14 should be the unlucky number. Yes, there were twelve "regular" seats for the gods at Valhalla, but there was a "high-seat" for the supreme Odin, and there is no mention of 13 (or fourteen, for that matter) in the legend itself. There isn't even any evidence that this supposed ancient superstition predated Christianity. Lachenmeyer says that the first recorded source of the Baldur myth is in the *Prose Edda,* written in the fifteenth century, "two centuries *after* the conversion of Iceland to Christianity."

And there are just as many holes in the Last Supper theory. Nowhere in the accounts of the betrayal of Christ is the number 13 mentioned, while twelve is mentioned several times. Lachenmeyer also argues that the twelve Apostles and Jesus had many meals to-gether (so why weren't the others unlucky?) and that it is

> inconceivable that the New Testament's authors would
> have wittingly embraced the blasphemy of implying that a
> group that included Jesus Christ—the son of God, the savior
> of man—was unlucky.

On the contrary, Lachenmeyer contends that 13 had positive connotations for Christians,

> precisely *because* of its association with Christ and his
> twelve disciples. To the Christian, 13 represented the benevo-
> lent 13 of Christ and his disciples in general, not the fateful
> 13 of the Last Supper.

Lachenmeyer lists many examples of prominent Christian theologians, such as St. Augustine, invoking the number 13 positively.

Another problem with tracing the ancient roots of triskaideka-phobia is that there is no written record of a fear of 13 before the second half of the seventeenth century, in England, when the notion developed that it was unsafe if 13 people sat at a table (often expressed as the fear that one of the 13 would die within a year). Lachenmeyer attributes the fright to the Great Plague of 1665, and the genuine panic caused by London losing nearly 15 percent of its citizens to the epidemic.

The European fear of 13 sitting at a table crested in the nineteenth century, when triskaidekaphobia mutated into a general fear of 13, but it was slow to migrate to the United States, which had positive associations with the number 13, because of the original 13 states. The 13-gun salute was the norm at patriotic gatherings in the

DAVID FELDMAN

United States, eventually yielding to the 21-gun salute (the origins of which we discussed in *Why Do Clocks Run Clockwise?*), but only after Vermont was added to the Union as the fourteenth state.

Considering the separate superstitions about Friday and 13, it's surprising that there is no recorded evidence of any special fear of Friday the thirteenth until the twentieth century. Lachenmeyer traces the fear of Friday in the United States to the New Testament and the Crucifixion, although he notes that Friday was also the traditional day for executions in the United States.

But what spawned the growth of the new fear? There is no smoking gun answer. Newspapers started taking note when Good Friday landed on the thirteenth in the early twentieth century, an indication that the superstition was gaining currency by the first decade of the last millennium.

But one huge event occurred in 1907. Thomas W. Lawson published a novel, *Friday, the Thirteenth*. As Lachenmeyer writes:

> it was this novel that redefined the coincidence of unlucky Friday and the 13th as one superstition, and launched Friday the 13th in the popular imagination. Lawson kept the superstition front and center from the opening sentence . . . to its dramatic conclusion. . . . [W]ith a plot that hinged on a speculator's attempt to manipulate the market on Friday the 13th, *Friday, the Thirteenth* was as successful as it was awful.

And Lawson's success did not end with a print bestseller. In 1916, a feature length silent movie version of *Friday, the Thirteenth* was released, furthering the superstition's grip. Sixty-four years later, Jason Voorhees carries on the tradition of trying to scare the dickens out of us with the first of the *Friday the Thirteenth* movies, even as the grip of the superstition withered.

So we buy the notion that a combination of the Last Supper, Good Friday, and Thomas Lawson is responsible for triskaidekaphobia, but it's important to remember a point that Lachenmeyer makes in *13*, perhaps the main reason we were reluctant to tackle this Imponderable until we read his book. Most of the books about superstitions were cavalier about ascribing the fear of 13 to one particular cause, and discussed the superstition as if it had not been mutated by different times and cultures:

> However, continuity of belief needs to be proved, not assumed. This is all the more critical in the case of number superstitions because numerology has been so widely practiced in so many cultures throughout history that it is difficult to find a number between 1 and 24 that has not been considered unlucky by more than one culture.

Exactly. You have to be methodical and analytical to untangle the messiness of irrational thinking.

Submitted by Mark Carroll of Nashville, Tennessee. Thanks also to Scott Comstock of Leavenworth, Kansas; Rose Marie Mielke of parts unknown; Patrick M. Premo of Allegany, New York; Pat Ryan of Churchville, New York; Destiny Montague of Peachtree, Georgia; Steve Brunton of Orlando, Florida; Wayne Goode of Madison, Alabama; Robert Bredt, via the Internet; and Lance Tock of Brooklyn Park, Minnesota.

DAVID FELDMAN

Since Doughnut Holes Are So Popular, Why Can't We Buy Bagel Holes?

n *Why Do Clocks Run Clockwise?*, we recalled the legend of Captain Hanson Gregory, who "invented" the doughnut—and the doughnut hole—when he impaled a solid fried cake on the spokes of the steering wheel of his ship. If Homer Simpson were a reader of ours, we know that his Imponderable would be: "What happened to that perfectly good doughnut hole that got punched out! And can I eat it?"

Bagels might not be as sweet as doughnuts, and reader Nora Corrigan may be no Homer Simpson, but as a bagel lover, she wonders why you can go into a Dunkin' Donuts or Krispy Kreme store and buy doughnut holes, but bagel holes seem to be nowhere in sight. The answer lies in the different ways bagels and doughnuts are produced.

Doughnuts are cut from a continuous sheet of dough. Two rings—one that forms the outline and the other that creates the center holes—are cut into that sheet before the dough is fried. Bagels start as slightly irregular strings—a hot dog–like shape—and are wrapped around a mandrel, a metal bar that helps form the distinctive bagel shape. Although many bagels are still handmade, bagel-making machines have been used since the early twentieth century, and they merely automate the same method.

So if the bagel hole is not stamped out but surrounded by dough, does that mean bagel holes are difficult or impossible to produce? According to the American Institute of Baking's "Dr. Dough" Tom Lehmann,

> It would be entirely within the production capability of most bakeries to create "bagel holes" as small round-shaped pieces of bagel dough that are processed in the same manner as bagels are, but to the best of my knowledge this has not been done commercially.

Who wants to be a millionaire?

Submitted by Nora Corrigan of Reston, Virginia.

Is There Any Logic to the Pattern of Train Whistles? Why Do You Often Hear a Signal of Long-Long-Short-Long?

Are you sick of all the commercials on American radio? Don't want to spend the bucks for satellite radio? One alternative is to tune in to the 160 to 161 megahertz bands to listen to one of the ninety-six channels that railroads use for internal communications—not a lot of music, but no commercials.

But trains didn't always have radio. Dating back to the days of the steam engine, railroad crews relied on other forms of communication, including whistle, horn, flag, and lantern signals, to let personnel within a train crew transmit important information.

The signals were not completely uniform. We have charts from six different railroads detailing their signals, including Union Pacific,

Northern Pacific, Southern Pacific, and B&O Railroads, and the codes are similar but not identical. In each case, though, manuals make distinctions between short and long "toots." In every case, one short signal tells the brakeman to stop the train. Two long sounds say: "Release the brakes and proceed."

In the days of the steam engine, when a train stopped, it was the role of the flagman, who usually rode in the caboose, to leave his perch and walk behind the train to make sure no approaching train collided from the back. Many of the early whistle codes were methods for the engineer to communicate with the flagman. One short toot followed by three long ones asked the flagman to protect the rear of the train, and three longs followed by one short asked him to guard the front. Other signals prescribed from which direction the flagman should reenter the train.

Generally, short signals indicate an urgent action. Even today, a series of short signals (usually at least seven) is a warning for a miscreant to get off the tracks—a train is approaching! When a train is moving and three short signals are sounded, it means that the train should stop at the next station; if a train is stopped, three shorts usually signal to proceed backward.

There's a reason why the poser of this Imponderable asks about two long, one short, and one long: it's one of the signals that the average person is most likely to hear. This is a standard signal to indicate that the train is approaching a public crossing. Here's the verbatim explanation for this signal from the 1943 Southern Pacific rule book:

> Approaching public crossings at grade, tunnels, and obscure curves; to be commenced sufficiently in advance to afford ample warning, but not less than one-fourth mile before reaching a crossing, and prolonged or repeated until engine has passed over the crossing.

DAVID FELDMAN

Of course, the loud warnings can be a bone of contention between railroad workers and the communities where the crossings are located. Fitz, a retired locomotive engineer from Chicago, Illinois complains:

> Today, all the yuppies complain about the noise. Few of
> them understand that without railroads, they wouldn't have
> the houses that they own, and the city that they live in.

Railroad workers required other signals to communicate with each other if one of the workers, such as a signalman, was out of earshot, or if conditions were too noisy for a whistle to be heard. Railroads have long used color signals to communicate the most basic commands. These colors could be signaled by a flag or by a stationary lantern at night. Red, of course, meant to stop; green said to go, but with caution. A white light signified that the coast was clear to go, and blue flags or lanterns were placed where men were working. Electronic color signals can now transmit more complicated information to train crews.

We thought that the days of lantern and hand signals were long over, but this is not the case. Retired Burlington Northern Santa Fe locomotive engineer Charlie Tomlin told *Imponderables*:

> I personally prefer to use hand or lantern signals, because it
> gives one better control over the movement when there is a
> line of sight. In a busy railroad yard (such as Eola, in Illinois,
> where I worked a good deal), there could be several crews
> working on the same radio channel and there was the danger
> of not being heard or being "walked on," not to mention the
> yardmasters giving instructions to the crews via the same
> channels.

It was just so much easier to have a job brief and agree on hand or lantern signals, using the radio only in an emergency. When you are switching with a small cut of cars, it is senseless to use the radio when the crew members can see each other or a lantern. As an engineer, I know that the new people are trained in giving and passing on hand and lantern signals and that they are trained to "exaggerate" the signal. I always told the new guys to give "big" hand and lantern signals. It is well appreciated by most engineers.

Lantern signals are simple, but dramatic, and not a little reminiscent of the movements of flag drill teams at football games. "Stop" is transmitted by a horizontal swing across the track—in an arm movement that mimics the "no" shake of the head. And the "proceed" signal resembles a nod up and down, with the arm raised and lowered vertically. If a signalman swings his arms up and down in a circle at full arm's length, watch out: the train has parted, not departed!

Submitted by Lawrence Atkinson, via the Internet.

Why Does Dog Food Have To Smell So Awful?

We always hear about dogs' vaunted sense of smell. The olfactory area in a human is about one-half of a square inch; a dog's is twenty square inches. While humans tend to trust their senses of sight more, a dog evaluates food and other living things with its sense of smell.

And with this heightened ability, the dog chooses to eat stuff that smells like dog food? Maybe we olfactory ignoramuses cannot savor the scent that is kibble, like children who can't appreciate the bouquet of a fine Burgundy. But we're not buying that. We're going to have to agree to disagree with our canine buddies.

Pet owners tend to anthropomorphize our dogs, so it's surprising that designer dog foods haven't been developed to make masters want to compete with Fido for the grub, but the pet industry maintains that its focus is on what pleases the pet. Robert Wilbur, of the Pet Food Institute, explains:

> Pet foods (unlike most human foods) provide the sole diet of most pets and the product must be complete and balanced nutritionally. In addition, pet foods must also be appetizing and appeal to a pet's sense of smell and taste. This is known as palatability and is a source of competition among pet food manufacturers. Scent and flavor must appeal to the dog and may differ from what would appeal to us.

Ironically, dogs, who will eat just about anything lying on the street, also have sensitive stomachs. Lucille Kubichek, of the Chihuahua Club of America, notes that efforts to find a scent that humans would like could lead to health issues:

> Dog foods carry the odors of the ingredients of which they are composed. I doubt food odors could be neutralized without adding one or more chemicals, which probably would be harmful to the animal.

By American law, dog food need not be fit for human consumption, and ingredient labels can be difficult to decipher. For example, dogs love lamb, and many kibbles include "lamb meal." What the heck is lamb meal? It consists of dehydrated carcass, including muscle, bone, and internal organs. For humans who are skittish about finding a hair in their soup, it might be more than a little off-putting to find that lamb meal is often infested with wool and with starch from the inside of the lamb's stomach.

Fat is often the second ingredient listed on pet food nutrition

DAVID FELDMAN

labels, and this is often responsible for that awful dog food smell. In her book, *Food Pets Die For,* Ann N. Martin lambastes its quality:

> Fats give off a pungent odor that entices your pet to eat the garbage. These fats are sourced from restaurant grease. This oil is rancid and unfit for human consumption. One of the main sources of fat comes from the rendering plant. This is obtained from the tissues of mammals and/or poultry in the commercial process of rendering or extracting.

While the pet food industry and its critics, such as Ann N. Martin, wrangle about whether commercial pet food is dangerous to their health, most dogs seem to be quite content to quickly clean their plates. That makes us happy. The faster they eat the dog food, the sooner the smell goes away.

Submitted by Dotty Bailey of Decatur, Georgia.

In Track Events with Staggered Starts, Why Do the Outside Runners Cut to the Inside Immediately Rather Than a More Gradual, Straight Line?

In middle-distance track events, such as the 800-meter run, the athletes start the race in lanes. Because the runners in the outside lanes must travel a greater distance than those on the inside, the starting lines are staggered, with those closest to the inside farthest back at the start. At the "break point," usually right after a full turn and at the beginning of the straightaway, runners in the other lanes have the opportunity to break to the inside to save running distance.

But sharp-eyed reader Dov Rabinowitz wrote us:

> The outside runners always seem to start moving toward
> the inside at the beginning of a long straightaway, and by the

time they are about one-quarter to one-half of the way down the straightaway, they have moved completely to the inside of the track, so that the runners are nearly single file.

Rabinowitz theorizes that runners waste extra steps to break to the inside prematurely, and that a sharper turn wastes some of the runner's forward momentum.

We posed Dov's Imponderable to four full-time running coaches and even more runners. All of the coaches agreed with Greg McMillan, an Austin, Texas coach, runner and exercise scientist, who preaches simple geometry:

> The question is a good one and the situation drives many coaches crazy. Every track athlete is taught that once you are allowed to "break for the rail," the athlete should run at a gradual diagonal to the inside lane. Since the point where the athlete may break from running in lanes is usually after the first curve (as in the 800-meter race), the best thing to do is run in a straight line toward the rail at the far end of the back stretch (around the 200-meter mark on most tracks). This will create the shortest distance around the track.

Obviously runners want to win, and running extra meters is an obvious hindrance to that goal. So we asked the coach why they do so. McMillan thinks that the fly in the ointment is often psychological:

> [The straight-line approach] sounds easy and every athlete will agree with it. But in the real world, this guideline goes out the window when the race starts. Most athletes will agree that it's the "safety in the pack mentality." The runner wants to get near the competition and feels vulnerable out on the open

track. So while it doesn't make logical sense to make a drastic cut toward the rail, the emotions of the athlete often cause this to happen. Some runners are better at controlling this urge than others.

Coach Roy Benson, president of Running, Ltd., based in Atlanta, Georgia, has coached professionally for more than forty years, and says that most runners have strategy rather than mathematics on their minds:

> Those in the outer lanes are usually trying to cut off runners on their left and take the lead as soon as possible.

But you can find yourself in traffic problems if you stay on the inside, too. Dr. Gordon Edwards, a coach and runner from Charlotte, North Carolina, wrote about the tactical dangers of breaking to the inside too soon:

> It would be impossible and hazardous to cut immediately to the first lane as you might impede other runners or bump into them. If you watch distance races, many runners run in the second or third lanes at times, even on the curves, for strategic reasons. Yes, they will run farther doing this, but sometimes it is a necessary tactic.

One of the reasons why runners might not break to the extreme inside is so that they can draft behind the lead runner, just as NASCAR drivers "leech" on the lead car. Rather than risk clipping the heels of the lead runner, it can be safer to be on the side. McMillan says that drafting is especially effective on windy days, and the benefits of running in the slipstream of another runner can outweigh the extra few meters the racer on the outside must complete.

We couldn't entice any of the runners to admit that they scooted to the inside prematurely for psychological reasons. All of them understood the merits of the gradual drift to the inside, and several chided other runners for darting to the inside prematurely:

> Every coach I ever had hammered home the distance-saving value of taking a tangent rather than cutting in; there's plenty of time . . . to make a cut, if the traffic pattern dictates it, but the farther away from the break point you are, the better, particularly because so many runners have a sheep-like mentality and break all at once; let them stumble all over each other fighting for the rail.

But the runners were afraid of traffic problems. One used a roadway analogy:

> [This problem is] not dissimilar from merging onto a highway . . . Although the straight diagonal line is slightly shorter, a quick analysis of the runners you are merging with might make one decide to cut in a bit quicker to avoid a potential bump, or to wait a bit, then cut in, also to avoid someone. In theory, if everyone is exactly even when they all go to cut in, and they all take the straight diagonal line to the pole, well, then that's one big jam up!

Some runners, drafting be damned, prefer racing from the front. If the runner feels he is in the lead, but is on the outside, he might want to cut over "prematurely" to get to the inside immediately and force competitors to try to pass him on the outside. But runners realize that it is possible to be "boxed-in," too—stuck in the inside lane, in a pack, with runners in front and outside of them, preventing acceleration.

Many a horse race has been lost because the jockey couldn't keep his mount from veering wide. Peter Sherry, a coach and former medalist at the World University Games, disagrees somewhat with the premise of the Imponderable. He thinks that the incidence of "elite" runners cutting to the left prematurely is lower than we're implying, and that it's more typical of high school races or races with less experienced athletes. He argues that if an elite runner darts to the inside quickly, there's usually a good reason—usually that

> a runner wants to make sure he gets a position on the rail before the race gets to the first turn. If you get caught on the outside of the pack during the turn, you will be running farther than someone on the inside rail.

Submitted by Dov Rabinowitz of Jerusalem, Israel.

For more information about the configuration of running tracks in general and staggered starts in particular, go to http://www.trackinfo .org/marks.html.

DAVID FELDMAN

Are Brussels Sprouts Really from Brussels?

Oh ye of little faith! You might not be able to find Russian dressing in Moscow or French dressing in Paris, but Belgium is happy to stake its claim on its vegetable discoveries—endive and Brussels sprouts and, for that matter, "French" fries.

Although there have been scattered reports about Brussels sprouts first being grown in Italy during Roman times, the first confirmed sighting of cultivation is near Brussels in the late sixteenth century. We don't know whether the rumor is true that there was a sudden upsurge of little Belgian children running away from home during that era.

We do know that by the end of the nineteenth century, Brussels sprouts had been introduced across the European continent, and had made the trek across the Atlantic to the United States. Today, most Brussels sprouts served in North America come from California.

Both the French (*choux de Bruxelles*) and the Italians (*cavollini di Bruselle*) give credit to Brussels for the vegetable, and Belgians are so modest about their contributions to the deliciousness of chocolate, mussels, and beer, who are we to argue?

Submitted by Bill Thayer of Owings Mills, Maryland. Thanks also to Mary Knatterud of St. Paul, Minnesota.

How Did They Mark Years Before the Birth of Christ? And How Did They Mark Years in Non-Christian Civilizations?

R eader Tim Goral writes:

We live in 2004 a.d. [well, we did a few years ago], or *anno domini* (the year of our Lord); the time period of human history that supposedly began with the birth of Jesus Christ. The historical references before that are all "b.c." (or sometimes "b.c.e."), so my question is: How did the people that lived, say 2500 years ago, mark the years? I understand that we count backward, as it were, from 2 b.c. to 150 b.c. to 400 b.c., etc., but the people that lived in that time couldn't have used that same method. They couldn't have known they were doing a countdown to one. What did they do? For example, how did Aristotle keep track of years?

In his essay, "Countdown to the Beginning of Time-Keeping," Colgate University professor Robert Garland summarized this question succinctly: "Every ancient society had its own idiosyncratic system for reckoning the years." We'll put it equally succinctly and less compassionately: "What a mess!"

Since we couldn't contact ancient time-keepers (a good past-life channeler is hard to find), this is a rare Imponderable for which we were forced to rely on books. We can't possibly cover all the schemes to mark time that were used, so if you desire in-depth discussions of the issue, we'll mention some of our favorite sources at the end of this chapter.

Our calendar is a gift from the Romans, but because early reckonings were based on incorrect assessments of the lunar cycles, our system has changed many times. "a.d." is short for *Anno Domini Nostri Iesu Christi* ("in the year of our lord Jesus Christ"); the years before that are designated "b.c." (before Christ). Obviously, the notion of fixing a calendar around Jesus did not occur immediately on his birth. Religious scholars wrestled with how to fix the calendar for many centuries.

In the early third century a.d., Palestinian Christian historian Sextus Julius Africanus attempted to fix the date of creation (he put it at what we would call 4499 b.c., but he had not yet thought of the b.c./a.d. calendar designation). In the sixth century, Pope John I asked a Russian monk, Dionysius Exiguus, to fix the dates of Easter, which had been celebrated on varying dates. Exiguus, working from erroneous assumptions and performing errors in calculation, was the person who not only set up our b.c./a.d. system, but helped cement December 25 as Christmas Day (for a brief examination of all of the monk's mistakes see http://www.westarinstitute.org/Periodicals/4R_Articles/Dionysius/dionysius.html). Two centuries later, Bede, an English monk later known as Saint Bede the Venerable, popularized Exiguus's notions. Christians were attempting to codify the dates of the major religious holidays, partly to compete with Roman and Greek gods and the Jewish holidays, but also to make the case for a historical Jesus.

Although the world's dating schemes are all over the map (pun intended), most can be attributed to one of three strategies:

1. *Historical Dating.* Christian calendars were derived from the calendars created by the Roman Empire. The early Romans counted the years from the supposed founding of Rome (*ab urbe condita*), which they calculated as what we would call 753 b.c. The ancient Greeks attempted to establish a common dating system in the third century b.c., by assigning dates based on the sequence of the Olympiads, which some Greek historians dated back as far as 776 b.c.

2. *Regnal Dating.* If you were the monarch, you had artistic control over the calendar in most parts of the world. In the ancient Babylonian, Roman, and Egyptian empires, for example, the first year of a king's rule was called year one. When a new emperor took the throne—bang!—up popped a new year one. Although Chinese historians kept impeccable records of the reign of emperors, dating back to what we would call the eighth century b.c., they similarly reset to year one at the beginning of each new reign. In ancient times, the Japanese sometimes used the same regnal scheme, but other times dated

back to the reign of the first emperor, Jimmu, in 660 **b.c.**

3. Religious Dating. Not surprisingly, Christians were not the only religious group to base their numbering systems on pivotal religious events. Muslims used hegira, when Mohammed fled from Mecca to Medina in 622 **a.d.** to escape religious persecution, to mark the starting point of their calendar. In Cambodia and Thailand, years were numbered from the date of Buddha's death. Hindus start their calendar from the birth of Brahma.

Looking over the various numbering schemes, you can't help but notice how parochial most calendar making was in the ancient world. Even scholars who were trying to determine dates based on astronomical events often ended up having to bow to political or religious pressure. And modern society is not immune to these outside forces—some traditionalist Japanese activists are trying to reintroduce a dating system based on the emperors' reigns.

Submitted by Tim Goral of Danbury, Connecticut.

For more information about this subject, one of the best online sources can be found at http://webexhibits.org/calendars/index .html. Some of the books we consulted include *Anno Domini*: *The Origins of the Christian Era* by George Declercq; *Countdown to the Beginning of Time-Keeping* by Robert Garland, and maybe best of all, the *Encyclopaedia Britannica* section on "calendar."

Why Are There More Windows Than Rows in Commercial Airlines? Why Aren't the Windows Aligned with the Rows of Seats?

When we posed this Imponderable to reader Ken Giesbers, a Boeing employee, he wrote:

> I understand this one on a personal level. On one of the few occasions that I flew as a child, I had the good fortune of getting a window seat, but the misfortune of sitting in one that lined up with solid fuselage between two windows. Other passengers could look directly out their own window, but not me. I would crane my neck to see ahead or behind, but the

view was less than satisfying. Why would designers arrange the windows in this way?

From the airplane maker's point of view, the goal is clearly, as a Boeing representative wrote *Imponderables,* to

provide as many windows as they reasonably can, without compromising the integrity of a cabin that must safely withstand thousands of cycles of pressurization and depressurization.

But the agenda of the airlines is a little different. Give passengers higher "seat pitch" (the distance between rows of seats) and you have more contented passengers. Reduce the seat pitch and you increase revenue if you can sell more seats on that flight. In 2000, American Airlines actually reduced the number of rows in its aircraft, increasing legroom and boasting of "More Room Throughout Coach" in its ads. When the airline started losing money, it panicked and cut out the legroom by putting in more rows.

In 2006, when soaring fuel prices are squeezing the airlines' costs, there isn't much incentive to offer more legroom. Their loads (percentage of available seats sold) are high; if they reduce the number of seats available on a flight, there is a good chance they have lost potential revenue on that flight. And passengers who are concerned with legroom might spring for more lucrative seats in business or first class. On the other hand, the more you squish your passengers, the more likely you are to lose your customer to another airline who offers higher seat pitch (the downsizing of American Airline's legroom in coach lost them one *Imponderable* author to roomier JetBlue). Web sites such as SeatGuru.com pinpoint the exact pitch dimensions of each aircraft on many of the biggest airlines, heating up the "pitch war."

A few inches of legroom can have huge consequences. Tens of millions of dollars can be involved in such decisions. Cabin seats can be moved forward or backward on rails fairly easily, but the costs are not just for equipment and the labor involved in reconfiguring the planes, but in downtime while they are reconfigured.

As you might guess, the sight lines of window passengers aren't uppermost in the minds of the airlines. If you happen to have a clear shot at the window from your seat, consider it serendipity.

Submitted by Alex Ros of New York, New York. Thanks also to John Lai of Billerica, Massachusetts; and Kevin Bourillon, of parts unknown.

How Do Audio Cassette Decks "Know" When To Go into Auto-Reverse?

Now that cassette decks in automobiles have been largely replaced by CD players and satellite radios, it's easy to forget that the auto-reverse function on cassette players was introduced in cars to allow drivers to keep their hands on the steering wheel instead of the stereo. The auto-reverse feature is reliable because it is so simple. It turns out that the reverse kicks in when a sensor "knows" that the sprockets (the two protrusions in which you click in the holes of the cassette) have stopped turning.

Robert Fontana, formerly of TDK Electronics Corporation, wrote to *Imponderables:*

> Auto-reverse cassette decks employ a special playback head
> and mechanical gears that are governed and switched electron-

ically. When a cassette tape reaches the end of side A, a sensor detects that the cassette hub is no longer revolving. Consequently, a signal is sent to reverse the polarity of both the capstan shaft and reel drive motors so that the cassette tape will travel in a right to left path. Additionally, the head gaps of the playback head are also switched to properly reproduce the left and right channels of side B. At the conclusion of side B, every parameter is then switched again to playback side A.

Submitted by Ernie Capobianco of Dallas, Texas.

Do Real Artists Line Up the Object To Be Painted by Putting Up the Thumb of Their Outstretched Arm? If So, Why?

The cliché: Tortured artist, clad in black, paces the room and finally stands in contemplation of his lovely muse. His unruly hair tamed by a beret, with palette in left hand, he surveys his subject with his right arm outstretched, gazing intently at his subject with his right thumb directly in front of him. The gesture seems simultaneously useless and pretentious.

But the experts we consulted gave a thumbs-up to using the thumb, or a reasonable facsimile (most often a pencil tip or a paintbrush) in exactly this fashion. In her essay, "Proportions in Figure Drawing," (found at http://drawsketch.about.com/cs/drawinglessons/

a/drawingintro.htm), About.com's guide to drawing and sketching, Australian artist and teacher Helen South, explains that the out-stretched arm trick is all about determining the proper proportions. For example, an artist will measure the height of a subject by lining it up with a pencil point. The average person is approximately seven and one-half heads tall (including the head). Here's how South in-structs the artist to use this technique:

> Remember that the basic unit in figure drawing is the model's head, from top to chin. Holding your pencil in a fist with the thumb upwards, and arm stretched out fully, close your non-master eye and align the top of your pencil with the top of the model's head, and slide your thumb down the pen-cil until it aligns with the model's chin. There you have the basic unit of measurement on the pencil. Repeat this step whenever necessary.
>
> Now, to find how out how many heads tall your model is, drop your hand slightly so that the top of the pencil is at the chin. Observe carefully the point on the figure that aligns with your thumb—this should be roughly below the breast-bone (two heads—you count the head itself). Drop the top of the pencil to that point, and so on, down to the feet.

Using this rough measurement, artists can draw horizontal lines on the canvas or sketchpad and have a rough blueprint of the appropri-ate proportions, especially if the artist is careful to always measure from the same point, with the arm totally outstretched, using the same instrument each time.

The artists we consulted confirmed that this technique pro-vides a useful reality check, not only in accurately depicting the size of a subject, but the angles and proportions of different elements in a composition as well. New York artist and graphic designer Joe Giordano wrote us about his approach:

> You close one eye to reduce everything to two dimensions and then use the thumb, brush, or pencil to line things up. For example, when drawing from a model in some exotic pose, you can see that an ear is in a direct line with the big toe and the knee lines up with the bent elbow, etc. Working from these points insures that your finished figure will be more or less in proportion since you are sometimes focusing in on details and are in danger of losing sight of the big picture.

Speaking of big pictures, we received a fascinating response from Sean Murtha, a muralist at the American Museum of Natural History. He draws gigantic dioramas for the museum (http://www.amnh.org/exhibitions/permanent/ocean/00_utilities/04b_murtha.php) and though he doesn't don a beret, he uses the classic technique, but innovates—Sean's a two-hander!

> What a question! I'm actually not a thumb user, but actually prefer my pencil or brush for the task, which I use primarily for establishing verticals and horizontals, rather than proportions. Since a rectangle usually defines a sketching surface, but no such geometry is present in the field, the pencil or brush held at arm's length introduces that line into one's field of view, making it easier to gauge the inclination of various edges and lines before you.
>
> Similarly, when establishing a composition, I use both my hands with thumbs extended at right angles to frame my view, blocking out everything but what will appear on the page or

DAVID FELDMAN

canvas. As for how I came to this technique, I have difficulty answering. All I know is that I've been doing it for a long time.

Detroit, Michigan artist Karen Anne Klein works on a much smaller scale. Many of her works are still lifes, often created with watercolor and pencil, which is awfully convenient—she has her two "measuring sticks" right at hand. Klein read Murtha's response and admits that for her, the technique is far from scientific:

> I don't use my thumb. I use my pencil or brush. Unlike Sean, I don't need to frame an image with my hands, as I am usually drawing only one object at a time and the creation of the composition comes from my head and not from the world. Since I am usually working on a single object, the use of the pencil is for determining either proportion or the degree of diagonal lines. Sometimes when I am combining objects that are in cases at the museum, I will use the pencil to compare sizes, but I rather doubt that it actually works since I am moving around. Who knows if I am really at the same distance from the second object?

Redlands, California artist Ruth Bavetta wouldn't put up with our guff about using a thumb or a pencil as a measuring device. We asked her, "Why not just draw a sketch and see if it coincides with your impression of the subject?" We gave her an example: What if she had to paint a picture of a girl and a ladder? Bavetta insists that no ruler is necessary:

> It's the ratio of sizes that's important. *But not the ratio of the actual sizes.* You don't want to know if the ladder is *actually*

taller than the kid, but if it *looks* taller from the viewpoint you're preparing to draw from.

You're not necessarily drawing by the scale of your thumb. You're just using it as an informal measuring stick to measure the relative sizes of what you see. Which *looks* taller: the ladder or the kid? And how much taller? Twice as tall? Three times as tall?

It is no more accurate than saying the ladder is 50 percent taller than the child. But what measurement are you talking about when you say that? The ladder may well *be* 50 percent taller than the kid, but if it's far away, and the kid is close, the kid will *look* taller.

The most important lesson in realistic drawing is learning to draw what you *see* instead of what you *know*. It's easy to be fooled by what you know. Because you know that the ladder is really taller than the kid, unless you make some kind of informal measurement where you stand, you're probably going to draw that ladder way too big.

Submitted by Nicholas Dollak of Fair Haven, New Jersey.

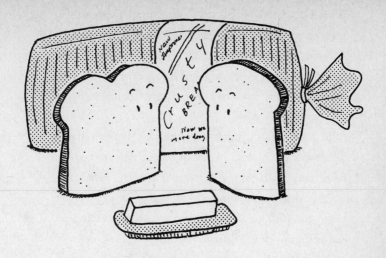

Why Does Store-Bought Bread Often Have Two Mounds on Top with a Channel in Between?

Perhaps the majority of prepackaged breads have flat tops, but like reader David Brandt we've often wondered why the tops of many loaves look like two arches with a gutter in between. When we have a question about bread or pastries, our thoughts drift to the American Institute of Baking, and when we think of the AIB, we contact Tom Lehmann, the director of bakery assistance who has acquired the nickname, "The Dough Doctor" in the baking and pizza industry. The good doctor swatted this Imponderable aside with aplomb:

> I think you are making reference to what the industry calls "split top" bread. This is a type of open top, or round top

loaf bread that is split down the center just before it goes to the oven for baking. Small retail bakeries often split the dough using a sharp knife or razor blade, but in large automated bakeries, it is accomplished through the use of a fine pressurized water stream, which cuts into the still soft dough as it moves to the oven.

In some cases, the water will contain a butter flavoring or even butter oil. In this case, the bread might be referred to as butter crust, butter top, or butter split bread . . . Some production lines that produce split top bread with a name like butter crust will use a type of mechanical blade to split the top of the dough and then immediately afterward, a small amount of butter oil is applied to the split. This results in a darker color at the site of the split and a buttery flavor.

The splitting of the top is [usually] done mostly for appearance purposes. For the most part, consumers have come to associate split top breads with a higher or premium quality bread.

Lucinda Ayers, vice president of Campbell's Kitchen, confirms Dr. Dough's explanation. In the varieties that feature split tops, Campbell's Pepperidge Farm bakers create the channel that runs from end to end down the top of a loaf by the process that Ayers calls "water cutting":

After the bread dough is in the pan and has gone through the proofing (rising) stage, the dough is puffed up very soft and high. At this point—just before baking—we run the puffed up dough through a stiff stream of water to "water cut" the channel down the back and give the loaf its distinctive shape.

DAVID FELDMAN

Why bother? Ayers says the reason for the process is "purely aesthetic—it gives the bread a more attractive, home-baked look."

When we first broached experts with this Imponderable, our own inability to describe the bread tops inadvertently led us to learn more about other types of lumps you see on the top of bread. For example, Kirk O'Donnell, vice president of education at the American Institute of Baking, told *Imponderables* that some bread dough is twisted before it is placed into the baking pan. When the dough rises, the twisting causes lumps. The twisting, according to O'Donnell, "improves the grain and texture of the bread." Many artisanal bakers don't *want* a smooth looking exterior, as upscale customers appreciate the imperfections if the bread exhibits the "homeliness of homemade."

Even more common, lumping is achieved through "docking," which is the deliberate cutting of the dough just before baking. Like water cutting, the main purpose of docking is aesthetic. O'Donnell explains:

> The reason for docking is to control the expansion of the
> bread in the oven. Most artisan bread is baked without a pan,
> so the bread tends to "burst" as it expands in the oven. By
> docking, the bread can expand evenly without bursting.

Pies and pizzas are usually docked, too, ordinarily by piercing the dough with the tines of a fork in several places. Without docking, bakers risk the chance of fissures developing in the finished product, which can lead to unhappy customers. It isn't easy to alienate a pie buyer, but it can be done.

Submitted by David Brandt, somewhere in Pennsylvania, via the Internet.

Why Does Water Vapor in the Sky Clump into Clouds Instead of Diffusing Evenly Throughout the Atmosphere?

Certain conditions are required for clouds to emerge. Most important, clouds form only in air that is saturated with ice or water. Most areas of the sky are neither moist enough nor cold enough to reach this saturation level, much to the pleasure of suntan lotion marketers. Steve Corfidi, of the National Weather Service, wrote to *Imponderables* emphasizing that clouds are only seen where air is forcibly uplifted and/or otherwise cooled to reach saturation. When air rises, it cools and loses its ability to hold moisture, leading to cloud and droplet formation. On the contrary, when air sinks, it warms, enabling it to hold more moisture, and droplets (and clouds) can evaporate back into invisible water vapor.

And how is this saturation achieved? Corfidi explains:

There are many ways by which air may undergo lifting and/or cooling to form clouds. For example, the cooling of moist air collecting in low spots at night can result in saturation and the development of patches of fog. Differential heating of the ground by the sun (e.g., strong heating over urban rooftops versus columns of rising air, called thermals) [can cause clouds]. If these columns extend high enough, condensation occurs and a cumulus cloud is born.

On an even smaller scale, small areas of upward and downward motion sometimes develop within an existing sheet of clouds. This most commonly occurs when there is shear in the cloud layer (winds which change in direction and/or speed with height) and the layer is thin. The resulting upward and downward motions produce the familiar dappled cloud pattern known as "mackerel sky."

Major contributors to cloud formation are "condensation nuclei," sites where water droplets form (clouds are nothing but the formation of millions of tiny water droplets and ice crystals— condensed water in liquid or solid form). Dust, pollutants, sea salts, volcanic ash, and residue from grass and forest fires can all be contributors, according to Boston meteorologist Todd Glickman, known for his weather reporting on WCBS-AM radio in New York. Glickman notes that the formation and dissipation of most clouds is a "very localized event."

The premise of this Imponderable implies that the sky is a worldwide, fluid whole, but Glickman urges us to think small:

Imagine pouring a quart of red food coloring into the Atlantic Ocean near a New York City beach. The coloring will not spread out and uniformly color the water all the way from

Iceland to western Africa to eastern Brazil. Rather, it will stay fairly concentrated in a local area, and be acted upon by waves, thermals, and other matters.

The atmosphere acts similarly; clouds will grow and dissipate in their own little space, independent of what is happening on the other side of the world.

Submitted by Randy K. Laist of Orange, Connecticut.

Your Souvenir of Club El So-da

Dec · 45

Why Is Carbonated Water Called *Club* Soda?

n various parts of the world, underground springs produce naturally carbonated water (but only where water has absorbed carbon
dioxide under high pressure). Some insist that effervescent spring
water aids digestion and helps cure various ills.

An Englishman, Joseph Priestley, who in his spare time accomplished another minor achievement (discovering oxygen), was the first
person to artificially create carbonated water. In 1772, Priestley published *Impregnating Water with Fixed Air,* to trumpet his achievement
(if you search the Internet, you will find several Web sites devoted to
the science and philosophy of this interesting character). Although
Priestley shared samples of his discovery with friends, carbonated water only became a commercial product in the early nineteenth century

when a Yale chemistry professor, Benjamin Silliman, bottled and sold seltzer water. Seltzer was and is, simply, filtered water with carbon dioxide added.

Silliman may or may not have been preceded by a Swiss chemist, Jacob Schweppe, who like Priestley, at first gave away his handiwork, but unlike the Englishman, eventually started charging. Schweppe's main contribution to carbonated water technology was perfecting a bottle that would retain the bubbles and allowed for mass distribution. By the mid-nineteenth century, J. Schweppe & Co., the predecessor of today's soft drink giant Cadbury Schweppes, was a thriving business in Europe.

The word "soda" was associated with many beverages as early as the late eighteenth century, according to Gregg Stengel, of Dr Pepper/Seven Up, Inc., but there are many claimants to the "club." According to lexicographer Stuart Berg Flexner's *Listening to America,* the expression "country club" was coined in the United States in 1867. From that point on, the word "club" continued to gain in popularity, conjuring an image of exclusivity, refinement, and panache. Decades hence, everything from "nightclub" to "club sandwiches" was exalted by these associations of class.

At antique bottle collector and author Digger Odell's Web site, http://www.bottlebooks.com/Carbonated%20Beverages/carbonated_beverage_trademarks%201890-1919.htm, you can see photographs of soda and carbonated beverage trademarks from 1890 to 1919 that show how "club" had infiltrated soft drink marketing. A 1901 trademark was granted to Country Club Soda Company Corporation, but ironically, club soda was not part of its line of products. One of the most valuable brands of that era was Clicquot Club, which marketed an extremely popular ginger ale—and no club soda.

Gregg Stengel told *Imponderables* that

Schweppes called many of their waters soda. The first reference to "club soda" was when the W.G. Pegram Company sold its beverage firm in South Africa to Schweppes in the 1920s. Pegram had a beverage which was called "club soda," which added to Schweppes soda, table waters and tonics. It is believed the name came from the many social clubs that were prevalent during that time. These clubs had musical concerts and arranged cricket matches.

Canada stakes its claim in the club soda sweepstakes, too. Many resorts in the Canadian Rockies have naturally carbonated springs, and the expression "club soda" became popular there. In the minds of many North Americans, club soda is associated with Canada Dry, a product that was indeed created in Toronto by pharmacist John J. McLaughlin. At first, he sold carbonated water to soda fountains (which were then almost always found in pharmacies), but in siphon bottles, to be squirted into fruit juices, fruit extracts, and syrups. Further complicating the Canadian "club connection" was the success of Hiram Walker, the first distiller to stop selling in bulk from wooden barrels and brand its product in smaller bottles for individual use. In order to give his product a touch of class, Walker called his product "Club Whiskey." The success of Club Whiskey set off alarm bells in the U.S. liquor industry and xenophobia in Congress—mandating that Walker call his product "Canadian Club."

Marie Cavanagh, director of information services at the National Soft Drink Association, wrote *Imponderables* that she was surprised that there is so little information about the origins of club soda. She buys the theory that the soda was used as a mixer in social clubs, but she was unable to verify this information from her organization's reference books.

While we can't claim to have a definitive answer, we did stumble onto one label on http://www.bottlebooks.com that gave us pause. Although the trademark was not registered until 1906 (still long before Schweppes sold its first club soda), an Irish company first sold a product called Cantrell & Cochranes Super Carbonated Club Soda in 1877. Is this the first club soda?

Submitted by John Beton of Chicago, Illinois.

Why Are There Holes in the Deposit Envelopes for Automated Teller Machines?

Sometimes, a hole is not just a hole. Those openings in ATM envelopes are known as "security holes" and they serve a purpose beyond providing fresh air for your hard-earned cash or paycheck.

In most cases, the party who is being protected by the security holes is the bank that operates the ATM. Customers are known to make mistakes and criminals are known to commit frauds. Joseph Richardson, of Diebold, Incorporated, told *Imponderables:*

> The holes in the ATM deposit envelope are an attempt by financial institutions to mitigate the risk of consumer error or intentional fraud. All ATM deposits are "subject to verification"

of the content of the deposit envelope once it is opened. Without the holes, an empty envelope was only discovered once it was opened and disputes occurred about what happened to the contents. With holes, the ATM deposit handler can see that the envelope is empty and leave it intact, thus mitigating the opportunity for disputes.

Some banks ask us to mark the outside of the envelope, indicating whether we are depositing cash or checks; others provide deposit slips that we slip inside the envelope. Most banks count the money by putting the cash deposits and checks into separate piles. Especially in the early days of ATMs, customers were reluctant to deposit cash, fearing that a random bank clerk would abscond with their money and head directly to the beaches of Cancun.

Fear not—the banks are at least as paranoid as any customers. Standard procedure dictates that ATM coffers are opened in teams, with at least one supervisor and one clerk checking the contents of ATM deposits. The holes allow the employees to see whether cash is included. Even when patrons are asked to specify on the envelope whether cash is being deposited, a surprising number of bank customers, whether unintentionally or not, omit the cash or include less than stated. For obvious reasons, the banks are hardliners about such disputes, especially involving cash—otherwise, they would be sitting ducks for scammers.

The location of the security holes is not random, either. They must provide visual access to the contents of the envelope but stay out of the way, as Richardson explains:

> ATMs are designed to print information regarding the transaction on the deposit envelope. This information is often used in balancing the transaction. Holes are placed on the

DAVID FELDMAN

envelope so that they will not interfere with the print line. Thus, the number and position of holes needs to be sufficient to quickly determine whether or not the envelope contains a deposit and still not interfere with the deposit handling module in the ATM.

New ATM technology eliminates the need for any deposit envelope. Checks are imaged and read for processing and bills are counted and validated at the ATM!

Most of the big banks in the United States are experimenting with image readers now. While most bank customers love withdrawing money from ATMs, many are still reluctant to deposit money to a machine, especially cash. Image readers confirm the amount of cash or checks deposited quickly, and the customer agrees to the amount before the money is snarfed by the ATM. The chances of a dispute have been lessened considerably, as the receipt for the transaction will always match the amount of money that the image reader confirms.

Although these new-fangled readers are costly for banks to buy and install, the savings for them are not just in reducing human teller transactions; the biggest savings is in back office operations. With envelope deposits, employees must open the envelopes manually, sort the checks and cash, key in data, run the checks through encoders, and integrate the deposit into the customer's account balance. With the new machines, the transaction itself can be processed right away. Whether this quicker processing leads to the customer getting credit for the deposit faster, we have our doubts.

Submitted by Crystal Nie of El Cerrito, California. Thanks also to Greg Kligman of Montréal, Quebec.

Why Do Women's Hips Sway When They Walk? Is This Intentional?

Sways with a wiggle, with a wiggle when she walks
Sways with a wiggle when she walks . . .

"What is love? Five feet of heaven in a ponytail
The cutest ponytail that sways with a wiggle when she walks . . .
    "What Is Love," words and music by Les Pockriss and Paul J. Vance

As the effect of the female sway upon men dates back long before the paean above (which was a big hit for a group called the Playmates in 1959), we'd always assumed that women swayed with intention and premeditation. So we were surprised when this Imponderable was submitted by a female. As anyone who watches *America's Next Top Model* can attest, swaying can be taught. But as we have learned, it can also be a naturally occurring phenomenon.

Scientists have pondered the female hip sway, and all the experts we consulted agree that there are anatomical and biomechanical reasons why women's hips sway more than men when they are walking. For one, women's pelvises tilt more when their legs are moving. According to John F. Hertner, professor of biology at the University of Nebraska at Kearney, the tilting is caused by three factors:

1. The female pelvis is relatively wide; in comparison, the male's is taller and narrower.
2. In women, the socket of the pelvis, which accepts the femur (upper leg bone), faces a bit more toward the front than the male pelvis.
3. The outward flare of the iliac crest, the upper, protruding bone in the pelvis, is greater in women than men.

Tom Purvis, a registered physical therapist and cofounder of the Resistance Training Specialist program, feels that these differences play out in practice. In Purvis's opinion, there is a purely biomechanical explanation for women's hip sway. According to Purvis, it is always more efficient and stable to support movements in the direct line of our center of gravity. If all that were necessary to walk efficiently was the even distribution of weight, it would make more sense for us to walk like tightrope walkers, placing one foot directly in front of the other.

But in practice, we don't walk this way, because two little things (or in some cases big things) get in our way—our hips. If you walk like a tightrope walker, you have to lift each foot up and move it around your body to get to the "rope"—it is time-consuming and awkward. But even if you look foolish "walking the line," your hips won't sway.

You could also choose to walk with your legs farther apart, directly beneath each hip. The result? Your entire body weight would shift from foot to foot, causing your whole body to sway from side to side. It's a cute way to locomote—penguins have been attracting other penguins by moving like this since waddling began. But for some strange reason, humans don't see waddling as being as sexy as swaying.

Most humans compromise by walking with the legs less far apart, and moving their centers of gravity, the hips, laterally over each foot when walking. Purvis concludes that women sway more than men because they have larger, wider hips, their feet are farther apart, and their center of gravity is lower to the ground. In order for women to walk, women's hips must move farther to the side over each foot. If Purvis's theory is true, then women with slimmer hips should tend not to sway as much and these biological verities explain why.

But not only biology determines destiny. Some women we've talked to attribute their intermittent swaying to one piece of technology—the high-heeled shoe. The higher the heel, the more the pelvis is thrust forward. Most women will sway more when wearing high heels, if only to try to remain vertical.

But not all swaying is unintentional. Throughout the animal kingdom, males and females tend to attract each other by accentuating the features that most distinguish their sex. As wider hips are a female trait, then it would make evolutionary sense for women to accentuate these sexual markers to attract potential mates, just as males might wear tight T-shirts to emphasize their muscles (and too often, their beer bellies, as well). Sex symbols, from Mae West to Marilyn Monroe to Jennifer Lopez have sashayed in front of drooling suitors. Of course, women are not immune to the lures of hip-swaying. In exaggerated form, the early Elvis ("The Pelvis") Presley outraged parents as he tantalized girls with his gyrations.

With the possible exception of dancers, no professional group is more associated with hip swaying than models. Willie Ninja, who works with models from OHM, IMG, and Metropolitan Modeling Agencies, specializes in developing models' runway walks. He talked to us about the vagaries of his unusual coaching career.

Ninja reports that women vary widely in the amount of natural sway. Some of the variation can be attributed to anatomy (torso length and bone structure, for example), but he emphasizes the importance of personality and cultural forces. Some women with natural movement will suppress the sway, attempting to deflect attention, while others will shake their moneymakers to ensure notice. He notes that some cultures, such as Latin and African, encourage and expect women's hips to sway noticeably, and models from these backgrounds generally do move their hips more. He has also noticed that women in rural areas tend to sway more than those in urban areas, the city dwellers presumably stifling their movements in order to deflect attention from aggressive males.

Ninja also believes that women's hip movements have not been immune to greater cultural forces, such as the feminist movement of the 1960s. As it became less acceptable for women to use their "feminine wiles," Ninja argues that women, at first consciously, and then later unconsciously, repressed the more blatant hip swaying of earlier times. Little girls then imitated their mothers, and the result is a generation of swayless women.

Nature has provided women with wider pelvic girdles than men, undoubtedly in order to facilitate childbirth. As in most cases of human dimorphism, the more prominent female attribute has been sexualized. In the sixteenth century, fashionable European women bought "hip cushions," worn underneath their skirts to double the size of their hips, or at least the appearance of their hips. And while we now seem to favor a thin waist, surely part of the appeal of the thin waistline is its ability to accentuate the wider pelvis and breast.

Although most women disdain corsets and girdles nowadays, Victorian women took their breaths when they could get them—when in public, they were often tethered into restraints that, if worn by animals, would attract the attention of PETA today. During times when more androgynous looks for females were in vogue (the flapper era in the 1920s, the Twiggy era in the 1960s), not only did women wear looser, less shape-defining clothes, but actresses and models seemed to deemphasize hip gyrations. It's hard to imagine Mae West slinking into a room wearing flappers, just as it would seem incongruous to see Twiggy, with her tomboy walk, wearing the kind of corseted outfits that Mae West or Pamela Anderson sport in their roles.

Even though women's anatomies might naturally promote more hip sway, it's hard not to believe that most swaying isn't attributable to cultural, personal, and intentional forces.

Submitted by Rachel Rhee of West Bloomfield, Michigan.

UNIMPONDERABLES: What Are Your Ten Most Frequently Asked Irritating Questions (FAIQ)?

Did you ever have a teacher who proclaimed that it is more important to ask good questions than provide good answers? We did. Unfortunately for us, he changed his position when we didn't do so well on our *test* answers.

So we'll understand it if you think us ungrateful wretches when we shout: Please, for our own sanity, please stop posing *faux*-Imponderables. After a hard day's night answering reader questions, we wake up at the crack of noon, craving new Imponderables. But too often we are faced with what we coined (in *Why Do Dogs Have Wet Noses?*) "Unimponderables," or "Frequently Asked Irritating Questions." In *Dogs,* we did answer a bunch of the questions that we did not consider real Imponderables, in what turned out to be futile attempt to stanch the flow. In *Pirates,* we will try again.

It's only fair for us to repeat the criteria we use to select the mysteries to answer in our books:

1. They must present genuine mysteries, something the average person might have thought about but that most people would not know the answer to.
2. The mysteries should deal with everyday life rather than esoteric, philosophical, or metaphysical questions.
3. They are "why" questions rather than who/what/where/when trivia questions that could be "Googled" in a flash.
4. Seminormal people might be interested in the questions and answers.
5. They are mysteries that aren't easy to find the answers to, especially from books. Therefore, they are questions that haven't been frequently written about.

How much simpler our lives would be if we could simply outlaw standup comedy! In *Why Do Dogs Have Wet Noses?*, we railed against Steven Wright, George Carlin, and Gallagher, who all pose "Why" questions and run off the stage, bathed in applause and laughter, before they have to answer any of them. We forgot to add Andy Rooney, who on *60 Minutes* can stand our hair on edge when he starts a sentence with: "Ever wonder why . . . ?" Before the Internet, at least the comedians' Unimponderables were confined to their fans—but now the fans spread these irritating questions around by forwarding massive lists of them to innocent victims.

Can we stop this scourge? Probably not. But in the spirit of goodwill, fellowship, and irritability, we offer our brief answers to our most frequently asked Unimponderables.

1. Why Do Drive-Up ATMs Have Braille Markings?

About five years ago, this dislodged "Why Do We Park on Driveways and Drive on Parkways?" as our most frequently asked irritating question, which was fine with us—at least this old warhorse involves more than wordplay.

Yes, it's true. If you look carefully at a drive-up ATM, you'll notice the same Braille markings as on the ones inside. Why do banks bother? The simple answer is: They have to. The Americans with Disabilities Act of 1990, revised in 2002, mandates that "Instructions and all information for use shall be made accessible to and independently usable by persons with vision impairments."

You might argue that any blind person who drives up to an ATM has more serious problems than vision impairment. But there are times when blind people do end up at drive-up ATMs. Obviously, the most common circumstance is when the blind person is driven by a friend, a relative, or a cab driver. Depending on the closeness of the relationship, the visually impaired person might not feel comfortable sharing a PIN number or the balance of an account with that other person. In some localities, especially on weekends in small towns, drive-up ATMs often become walk-up ATMs (not just by the visually impaired). The spirit of the ADA is that folks with disabilities, ranging from mobility to vision issues, should be able to accomplish as many tasks as possible independent of others.

We spoke to a group of Banc One executives in Denver, Colorado, who told us that they initially opposed this provision of the ADA in 1990, and were supported by their powerful trade group, the American Bankers Association. But when Banc One consulted with the ATM manufacturers, the executives realized that it would be more trouble than it was worth to use different configurations for ATMs anyway, and possibly more expensive.

Ironically, the Braille markings on walk-up and drive-up ATMs have come under fire from the visually impaired. Studies vary, but the consensus seems to be that approximately 15 percent of blind people are Braille-literate. Blind customers complain to banks when the ATM software changes, which necessitates learning a new set of keystrokes to complete the same transactions. As ATMs now allow customers to buy postage stamps, pay bills, look up their mortgage balance, and more, it increases the difficulty for blind people in performing even the simplest of transactions.

For these reasons, banks are being pressured to provide "talking ATMs," ones that not only offer the customer a menu of choices for each transaction, but vocalize the contents of each receipt that is printed. But there are problems, too, beyond the obvious expense in programming and equipment. If the ATM speaks out loud, then passersby could hear intimate financial details (a thief might be more than casually interested in the dollar amount of a withdrawal). If a headphone system is used, who supplies the headphone? If it's the responsibility of the bank, then how will the headphone be offered when the bank is closed? And must the bank supply headphones to drive-up ATMs?

Now you understand why Braille on drive-up ATMs might be an obsession to stand-up comics, but not the first concern of banks or the visually impaired.

2. Do Blind People Dream? If So, Do They Dream in Color?

They sure do dream, but most people who are blind at birth or become blind at an early age (up until the age of five or six) tend not to see anything in their dreams, although a small minority reports observing shadows or other abstract patterns. Most folks who were blinded at age seven or later experience visual images, sometimes just as vivid as fully sighted people, but often these images deteriorate over time. And yes, if they see images in dreams, they view them in color.

One of the best recent studies, "The Dreams of Blind Men and Women," conducted by psychologist Craig S. Hurovitz and three colleagues (and available online at http://psych.ucsc.edu/dreams/Articles/hurovitz_1999a.html) corroborates these generalities but also finds that blind folks who can see limited or no visual imagery not only hear in their dreams, but often experience sensations of touch, smell, and taste. As far as we know, however, blind people do not dream about drive-up ATMs.

3. Why, Unlike Other Sports, Do Baseball Managers Wear the Same Uniform as the Players?

As anyone who has watched *The Office* will attest, management isn't always all it's cracked up to be. Nowadays, the onfield leaders of sports teams try to market themselves as combinations of the best attributes of Albert Einstein, Machiavelli, Julius Caesar, and Mother Teresa in order to justify their huge paychecks. But in the early days of baseball, the game managed to survive without a manager at all.

Before the twentieth century, teams had player-captains, but as the salaries in baseball increased, teams encouraged their aging stars to don the mantle of manager. Many of them spent years as player-managers, and wore the standard uniform for the most obvious of reasons—they still played the game. Player-managers and player-coaches were common well into the mid–twentieth century, but the tradition of managers wearing player uniforms has continued to this day, regardless of their physical shape (who could forget Tommy Lasorda waddling to the mound in his Dodgers uniform?). The last player-manager in the big leagues was none other than Pete Rose, the Cincinnati Reds' troubled second baseman.

No law prohibits managers from wearing Armani if they wish, and a couple of Major League Baseball managers have bolted from tradition. Connie Mack, Hall of Fame leader of the old Philadelphia Athletics, who still owns the record for most wins by a manager, wore a suit to work. Burt Shotton, an ex-player who managed the Brooklyn Dodgers in the late 1940s, wore a team jacket over his suit and tie, a look that was unique, if unlikely to pass muster with the fashionistas.

We can't think of any other sports where on-the-field management wears the same outfit as the players. We know of only one example where this tradition was violated even temporarily. In hockey's 1928 Stanley Cup final game, Montréal Maroons player Nels Stewart fired a shot that hit New York Rangers goalie, Lorne Chabot, in the eye. Faced with no suitable replacement, 44-year-old Rangers coach Lester Patrick strapped on the goalie gear and played brilliantly—and the Rangers went on to win in overtime, 2–1.

4. How Many Licks Does It Take to Get To the Center of a Tootsie Pop?

Although we get flooded with this question, its genesis is not from a standup comedian but from a 1970 television commercial that launched a long-lasting campaign. We classify this conundrum as an Unimponderable for what seems like an obvious reason: All licks are not created equal—there can never be one answer to this question because there is too much variation in the way Tootsie Pops are consumed. All lickers are not created equal, and neither are the Pops themselves. Does the number of licks increase when the Tootsie Roll center is, as is often the case, off-center? To the best of our knowledge, there are no empirical studies. And how can you calibrate your research to account for the irrefutable tendency of lickers to suck on the Pop?

All of these obstacles have not prevented the curious from conducting scientific experiments to determine the "lickiocity" of Tootsie Pops. In at least two bastions of higher education, Purdue University and the University of Michigan, students have concocted machines to simulate the licking mannerisms of humans, and they actually came to similar results (364 and 411, respectively). The Purdue students enticed twenty human volunteers and found that they averaged 252 licks per Pop.

Are there regional differences in Pop licking? Based on the data collected by the high school students at the Mississippi School for Math and Science, we'd have to assume so. Students attempted to test their hypothesis that boys would get to the Tootsie Roll center of the Pop faster than girls and quickly confirmed it: the mean for girls was a whopping 1,656 licks, versus 1,239 for boys. But the number was so much higher for these high school students than the college students. Is there a direct correlation between age and licking power? Higher education and consumption? Not clear! For it took junior high students at the Swarthmore School an average of a mere 144 licks to get to the center, demonstrating precocity, or perhaps, merely starvation.

Judging by our mailbag, it isn't surprising that Tootsie Roll Industries claims that it has received tens of thousands of results from children, with an average of 600 to 800 licks. The discussion on its Web site indicates that Tootsie Roll has given up: "Based on the wide range of results from these scientific studies, it is clear that the world may never know how many licks it really takes to get to the Tootsie Roll center of a Tootsie Pop." And we can live without an answer. After all, we agree with Mr. Turtle, who in commercials was asked how many licks it took him to get to the center. He replied: "I never made it without biting."

5. How Does the Hair on Your Head and Chin Know To Grow Long, But the Hair on Your Eyebrows or Arms Stays Short?

Although Andy Rooney and other folks with bushy eyebrows might not agree with the premise, we get the idea. Next to the Braille/drive-up ATM mystery, this is our second most frequently posed Unimponderable. Not a bad question, but the answer to the part we can answer is readily available, and the rest can never be answered.

The hair we see on our head and body doesn't know much; in fact, by the time we can see it, it is quite dead. Inside each hair follicle, new cells form and push out older cells (the hair that we see) during the follicle's growth phase. But the follicle goes through a rest phase, as well, when the hair shaft breaks, and the new hair replaces it. The varying lengths of these rest and growth phases determine the length of our hair. None of this is synchronized, or we would shed like a Golden Retriever.

We are constantly losing hair, as our bath drains will attest—it is just that the growth phase for the hair on our head is longer than for other parts of the body. Why is that? Some biologists argue that the hair on top of our head was originally a survival mechanism to shield us from the sun. Most evolutionists argue that the main purpose of our manes is to attract the opposite sex (the folks at Clairol, Grecian Formula, and Rogaine subscribe to this theory), and that the beards and mustaches on men are signals—just in case women don't notice the bulging muscles and remote control in hand—that the fairer sex is looking at striking specimens of male pulchritude.

6. Why Do We Park on Driveways and Drive on Parkways?

We believe this Unimponderable sprang from the fertile mind of comedian Steven Wright. To the best of our knowledge, he's never answered it. Once again, we will milk his wordplay any vestige of wit, and answer this one more time if you vow on a stack of *Imponderables* books never to ask again. To quote from the scholarly yet illuminating classic, David Feldman's *Who Put the Butter in Butterfly?*:

> One of the main definitions of *way* is "a route or course that is or may be used to go from one place to another." New York's master builder Robert Moses dubbed his "route or course that was used to go from one place or another" *parkway* because it was lined with trees and lawns in an attempt to simulate the beauty of a park. The *driveway*, just as much as a *highway*, or a *parkway*, is a path for automobiles. The driveway is a path, a *way* between the street and a house or garage.

Imagine a time when the government was trying desperately to encourage us to drive more often—this was the inspiration behind the dubbing of Moses's "parkway."

7. Why Do They Use Sterilized Needles for Lethal Injections to Prisoners? Why Do They Use Rubbing Alcohol on Their Arms?

Some variation of these two questions seems to be on just about every e-mail forwarded "Things That Make You Go Hmm" list. Like the Braille-

ATM question, it probably started as a one-liner from a comedian, but it is so-often asked here at Imponderables Central that we'll answer it seriously on the assumption that some folks do want to know the answer. Actually, there are a whole bunch of reasons why executions are handled meticulously:

1. *Practical Advantages:* Even if the main purposes of an execution are deterrence and retribution, there are practical reasons for using sterilized needles. For one, needles come packaged sterilized to begin with. Imagine the public relations nightmare if the State injected prisoners with used or dirty needles. The purpose is to kill the prisoner, not torture him.

Likewise, the rubbing alcohol doesn't just disinfect the area to be pierced—it also raises the blood vessels closer to the surface of the skin, making it easier to find an entry point for the needle. Several executions have been delayed for up to an hour while technicians futilely tried to find a vein to inject. In a famous case, the execution of Randy Woolls in Texas in 1986, the condemned man actually helped technicians to find a usable vein.

One other obvious practical advantage to these precautions—they help protect the technicians administering the drugs. Why should they be subjected to less than sterile conditions?

2. *Political Reasons:* Opponents of capital punishment seize on any opportunity to criticize lethal injection (as well as other methods of legal executions), and the most obvious argument is that capital punishment is cruel and unusual. One of the reasons why lethal injection is now used in thirty-seven of the fifty states in the United States is that it is perceived as less cruel than hanging, firing squads, electrocution, or the gas chamber. The less humane the actual execution appears, the less support lethal injection will garner from citizens.

Opposition doesn't come only in the courts. The American Medical Association's Code of Ethics prohibits doctors from aiding in executions. With a few exceptions, physicians don't administer the lethal injection itself, although in some states a doctor is in attendance to at least monitor the proceedings. Most of the technicians who administer lethal injections are not experts in anesthesia, and as a result there have been many accidents in which the condemned prisoners suffered unexpected adverse reactions to the drugs (in most cases, three different drugs are administered). If only for political reasons, the state governments and capital punishment supporters don't want witnesses, especially the press, to describe the execution as torture.

3. Legal Reasons: Procedures for executions are mandated by each state, and the ones that we've seen include provisions for everything from the prisoner's last meal to the specifics of the administration of the lethal drugs. For example, the California procedures mandate that an IV line be attached to two usable veins (in case one line malfunctions), prescribe the exact amount of sodium pentothal, pancuronium bromide, and potassium chloride to be injected, insist that the line is flushed with normal saline solution in between the first and second injection, and require that a physician be present to declare the time of death.

4. Just in Case: And what if there was a stay of execution after the process of lethal injection has started? It actually happened in 1983 to James Autry, a Texas convict. Thirty-one minutes before his scheduled execution, with needles already inserted and saline solution running through his veins, the Supreme Court issued a stay (Autry was eventually executed the following year).

5. Psychological Reasons: Our comrade in Q&A-dom, Cecil Adams, wrote about this question in his "Straight Dope" column. After citing some arguments similar to the ones we mention above, he put forth his personal theory (which we liked):

> Which brings us to what I think is the real reason for swabbing the arm—it allows the executioners to think of themselves as professionals doing a job rather than killers.

Interviews with members of execution teams reveal that they place great stock in following proper procedures. We may be certain that if the prisoner were to choke on a chicken bone during his last meal, the authorities would spare no effort to save his life an hour prior to ending it.

Another reason to put stock in the psychological theory is that in many states, there are several "executioners," none of whom know which one administered the fatal dose, just as members of a firing squad weren't sure which of them actually killed the condemned prisoner.

8. Why Did Kamikaze Pilots Wear Helmets?

This Unimponderable, from the folks who brought you parkways and driveways and Braille-laden ATM machines, is on most Internet "Things That Make You Go Hmm" lists. Again, we have no idea who first uttered this question, but whoever it is, he or she should be getting royalties, for it is has captured the imagination of ponderers everywhere.

Some folks treat the question as a joke ("Was the helmet to cover up morning hair?") while others sincerely want to know the answer. Some plausible reasons that kamikazes might have worn helmets while flying their suicide missions:

1. In case of antiaircraft fire, the helmet might have provided some protection.
2. The helmet protected pilots from jostling in the plane, especially hitting their heads against the sides of the cockpit or the canopy of the plane.
3. Radio communication gear was inserted in the helmet. The electronics needed protection against being damaged during a ragged flight.
4. The helmet was there for warmth. It got cold in those cockpits!
5. The helmet was part of the uniform of a Japanese pilot. There were no special uniforms for kamikazes.
6. Many kamikaze missions were abandoned. The pilots might as well have remained as safe as possible if they didn't succeed in their mission.

All of these explanations are reasonable, but there's one tiny wrinkle in the equation: Kamikazes didn't wear helmets. They wore the same leather flight caps (and goggles) that were standard-issue equipment for other pilots. If you enter "kamikaze pilots" in Google Image Search, you'll see scores of images of kamikazes, most in full uniform with leather caps, and none with hard helmets. While doing research on this Unimponderable, we found a fascinating Web site, Kamikaze Images (http://wgordon.web .wesleyan.edu/kamikaze/index.htm), a project created by William D. Gordon for his master's degree at Wesleyan University. The site is full of images of kamikazes, and interviews with Japanese and Americans about the subject.

When contacted, Gordon was more than familiar with this Unimponderable, and tired of it, too ("Personally, I hope your book's explanation stops this joke."). Gordon thinks that the premise assumes that kamikaze pilots really wanted to live instead of carrying out their suicide mission. But Gordon says that based on his research, that is not the case. Kamikazes did not take on their task lightly, and were trained and prepared for their missions for many months.

Gordon confirms that the leather cap and goggles were "the same as any other pilot's in the Navy or Army." Many times kamikaze pilots wore a *hachimaki* (headband) with some saying written on it, "but I do not think this was that common for other pilots."

We asked how often kamikaze missions were abandoned, and what the primary reasons were. He responded:

Many flights did not result in a ship being hit for the following reasons:

- mission postponed due to weather
- unable to take off because of mechanical problems
- returned to base due to not finding ship or due to bad weather
- shot down before reaching a ship

- forced (or crash) landing somewhere other than sortie base due to mechanical problems or enemy attack
- ran out of fuel before finding ship
- missed ship (e.g., pilot error, plane on fire)

The records are not available or are inconsistent, so an accurate percentage is impossible no matter how you want to calculate the figure. A rough estimate of kamikaze planes that sortied and hit a ship is about ten to fifteen percent.

9. Why Are "Black Boxes" Orange?

If Army tanks can be painted in camouflage to hinder detection in battle situations, then why can't critical flight recorders be painted bright colors to ease finding them at disaster sites on the ground or in the ocean? The Federal Aviation Administration mandates not only that all large commercial aircraft must be equipped with two black boxes, but that they be painted "either bright orange or bright yellow." We haven't been able to find any examples of black "black boxes"—as far as we know, they are invariably orange.

That's right, there are *two* black boxes in each plane: a cockpit voice recorder that monitors pilot conversations with each other, air traffic control, and ground or cabin crew, as well as engine noises; and a flight data recorder, which according to the National Transportation Safety Board, monitors "at least eighty-eight important parameters such as time, altitude, airspeed, heading, and aircraft attitude." The most sophisticated black boxes now monitor close to 300 flight variables.

Both boxes are placed in the most "survivable" part of the aircraft, usually in the tail section. Each of these recorders are equipped with an underwater locator beacon, which sends an ultrasonic pulse so that receivers can locate them in the water, and are built to withstand water pressure as deep as 20,000 feet below sea level. To further assist their retrieval in usually difficult circumstances, the boxes are not only painted a bright color, but are covered with fluorescent reflective tape.

We've heard many different theories about why the recorders were called "black boxes" in the first place. As black boxes are sought after crashes, one theory is that "black" refers to the death surrounding their use. Boeing's Ken Giesbers remembers the term "black box" from engineering school to describe a device having unknown contents:

> When testing a black box unit, you could apply inputs and measure outputs, but you did not know what was going on inside the unit. Considering how rugged and well-sealed these devices are, I think they qualify as "black boxes" to many people.

Were black boxes actually ever painted black? So far, we haven't been able to verify that they were.

10. If You Can Answer the Questions in Your Book, Shouldn't They Be Called "Ponderables" Instead of "Imponderables"?

We frequently receive comments from readers arguing that if we can answer one of the questions in our books, then the conundrum can't be called "imponderable." If we answer a question, haven't we pondered it? Our *Merriam-Webster's Collegiate Dictionary* defines "ponderable" as "significant enough to be worth considering." Surely, no one could think that the consideration of why you never see baby pigeons could possibly be insignificant. If there are such folks, we hope they seek counseling for their wounded souls.

But to us, "imponderable" means more than just "not ponderable." It refers to the state of anxiety we feel when a question burns to be answered but we fear it never will.

Some dictionaries also define "imponderable" as a question that can't be answered by numbers or exact measurement. That's a description we can buy. We promise that the answer to an Imponderable will never be "15" or "True." The mysteries of everyday life usually defy easy answers, and if a question is interesting enough, who cares about its practical application? It's worth pondering.

Why Do the U.S. and Other Navies Use a *Fouled* Anchor as Their Symbol?

A fouled anchor is one that has its line or chain wrapped around it. No sailor worth his salt would show pride in mucking up an anchor in this way, yet the graphic depiction of the fouled anchor has graced the uniforms of the U.K., U.S., and Canadian navies for centuries.

The fouled anchor has an illustrious lineage. It was the official seal of Lord High Admiral Charles Lord Howard of Effingham, who oversaw the English victory over the Spanish Armada in 1588. Ever since, the fouled anchor remains the official seal of whoever occupies the seat of the Lord High Admiral of Great Britain.

We can find no answer to this Imponderable more convincing than the Naval Historical Center's. If any group would seek to justify the inappropriate imagery, it should be the Navy. And yet there is no better response than the Navy's: "It looks pretty."

It seems strange that the navies of the world should use as an insignia the abomination of all good sailors. Somewhere back in the early days, a draftsman with more artistic ability than technical knowledge produced the well-known design which shows an anchor with its cable hopelessly fouled around the shank and arms. How such a design could win the approval of the Admiralty Board is beyond comprehension, but the fact remains that the sight of the fouled anchor has become an international emblem.

Perhaps it could be argued that the fouled anchor is like the choking sign in a restaurant—a cautionary symbol. But we doubt it. Navy personnel have had a long tradition of getting a tattoo after they have completed their tour. The most popular design? A fouled anchor.

Submitted by Michael, via the Internet.

Why Are Computer Circuit Boards Green?

Before we get deluged with mail—we know! Yes, you can find circuit boards in almost as many colors as Crayola has crayons, including red, gold, pink, yellow, brown, and black. Tracy Elbert, product manager for PCBexpress, wrote *Imponderables* that her company used to manufacture rainbow-colored boards, with "a drizzle of all the colors spiraled around randomly." But dark green boards have become the default color. Why?

On one level, the answer is easy. The green color on most circuit boards, not just on computers, but on everything from pinball machines to television sets, comes from the "solder mask." The solder mask, found on both the top and bottom of a PCB (printed circuit board) can be applied as a dry film, printed in a silk screen–like process, or now most commonly, applied in a liquid form.

The purpose of the solder mask is not to look pretty but to cover and insulate the parts of the circuit that don't require soldering. All kinds of bad stuff can result if solder strays from its intended destination—shorts being the most serious consequence, as Mike Lopez, of PCB manufacturer Prototron Circuits, explains:

> The green color actually is an epoxy-based material that acts as a solder resist. When components are loaded onto a board, the solder mask resists the solder from jumping over to another hole or pad location, which would cause a short in the circuit. Now the assembly process becomes easier knowing that if you load all the components from one side and run it along with a conveyor belt and flowing solder beneath it, all of the leads that come through the holes will wick up perfectly and keep the solder to the metal only and not form a big solder mess.

We also know that there is a certain amount of inertia with the colors of things. If circuit boards have always been green, then why rock the boat? The manufacturers don't have much incentive to change, unless customers seek new colors. But Elbert points out that there is cleanup and waste involved in her plant switching back and forth between different colors—the dark green default is fine with her.

We know that the oldest circuit boards were not green. Bob Dietzel, who spent twenty-nine years at Bell Labs doing integrated circuit research and development, told *Imponderables* that before commercially applied coatings were available, Bell Labs made its own boards, which were translucent white.

> These boards were soldered by hand in those days so no mask was needed. We had a shop full of "wiremen" who placed the components on the boards and soldered them.

The earliest circuit boards were masked by hand with a tar-like substance and etched by hand in an acid solution. This soon was replaced with masks made with rub-on masking and etched with ferric chloride solution.

Tom Wagner, an audiovisual technician and Web designer from Malaspina, British Columbia wrote about the rainbow of colors found on early boards:

> The color of the board shows the composition of it, hence the use of it.
> Original and older boards were Bakelite composition, and were brown.
> Phenolic [boards were] . . . tan.
> Most "non filled" fiberglass boards used in computers are green.
> Filled fiberglass is blue.
> Teflon is white.

But why was the solder mask green in the first place? One theory, advanced in 2000 on IPC's TechNet E-Mail Forum by David Albin, of Coates Circuit Products, is that the solder mask manufacturers used the materials at hand, in this case a

> copper phthalocyanine pigment, which at the time of the first boards, was the only heat-stable organic colour [available].

Others "blame" Uncle Sam for our dark green solder mask. Mike Lopez writes:

I've heard that PCBs were starting to develop colors back
in World War II days and green was the color of choice for
the military.

Another poster on the TechNet board sides with the armed
forces theory:

The green color of solder mask was chosen after extensive
testing by the U.S. military at the National Materials and Pro-
curement Center in Cedar Bluffs, Virginia in late 1954. According to Colonel Robert Bright, Public Liaison Officer, the
particular shade of green was found to provide the maximum
contrast to the white silkscreen ink under all tested adverse situations while still allowing a clear view of the underlying circuitry.
Every other color tested failed to provide the required contrast
under conditions of low/high illumination levels and various colored light sources. The quantity of printed circuit boards purchased by the Defense Department established the "de facto"
standard which most vendors began to follow. Usage of other
colors is permitted for prototype and or pre-release boards.

Regardless of whether or not the military mandated dark green,
some argue that green is the color easiest on the eye, especially for long
periods, while providing sufficient contrast for board assemblers and
inspectors who were supervising their work. Even the manufacturers
of PCBs have a selfish reason to prefer green, as Tracy Elbert explains:

The liquid solder mask we use is like an epoxy, which
needs to be cured to harden. Green is the easiest color for the
UV light to accept, making the process easy for us.

But some manufacturers of electronics equipment color-code their PCBs when developing or testing new circuits, in order to distinguish between different kinds of boards, or to test the efficacy of one over the other. Paul Morgan, of circuit board manufacturer K&F Electronics, told *Imponderables* that solder masks are available in all the primary colors,

> though mixing any color is possible. Black and white are also available, making gray yet another choice. Some companies use the different colors for color coding, such as red for experimental research and development and green for production. Black is typically used to block light for applications such as automobile dash board displays.

Most big PCB manufacturers offer designer colors, sometimes at premium prices. But it will take more than a passing fad to make anything other than dark green the "new black" in the solder mask biz.

Submitted by Scott Bossak of Brooklyn, New York.

Why Do Wet Dogs Go Out of Their Way To Shake Off the Moisture in Front of Their Master?

Dogs can definitely be trained not to shake off water near people, but given their druthers, they'll do it every time. Fred Lanting, a dog judge, breeder, and author, notes that if dogs don't see anyone, or are wild or feral, they don't wait to shake off. But Lanting observes that when a domestic dog has his owner around, Fido wants to share the moment:

Why do dogs head for you and *then* shake the water off? Perhaps they think your name is Everest. Because, like Everest, you are there. If no one were near, the dog would shake as soon as convenient. But dogs are social animals with a very strong emotional attachment to and dependency upon man, and their psychological desire to be near you is momentarily stronger than their physiological need to shake off the water.

Dog trainer and author Suzanne Clothier concurs:

> In my own pack of six to eight dogs, I have noticed that when play is not involved (i.e., my dogs are swimming at free will in the pool or stream), they shake almost immediately upon exiting the water. When I or another person is involved, as we often are throwing bumpers or balls into the water, the dogs exit the water but delay shaking until within three to six feet from us.

Is there the tiniest chance that dogs might be exhibiting a little catlike streak? Robert Habel, professor emeritus of Cornell University's College of Veterinary Medicine ponders:

> Possibly they seek vengeance for being sent into the water after a stupid stick. Possibly they want immediate commendation for going into the water. Or for coming out.

Nah. Dogs don't have enough guile to splash us passive-aggressively. We side with the dog trainer JP Yousha, of El Paso Texas, who suspects that dogs shake in front of their masters

> to share their joy. A wet dog, at least one who enjoys the water, is in no hurry to shake and naturally returns to the one he adores to show his delight.

Submitted by Glenn Jones of Wantagh, New York.

Why Is a Square Boxing Area Called a "Ring"?

Even a punch-drunk fighter can figure out that his work space is a square. Where did "ring" come from?

The answer is that "ring" was first applied not to the boxing area, but to the spectators who formed a *ring* around the combatants, according to *Brewer's Dictionary of Phrase and Fable*. Although hand-to-hand combat was probably invented by the first two-year-old boy to discover he had a younger sibling, the first public boxing matches took place in early eighteenth-century England. These were bare-knuckled, no-holds-barred affairs with no time limits, no ropes, and no referees. The winner was the last man standing. The ring of bloodthirsty fans formed a permeable enclosure for the pugilists.

Eventually, as boxers started to make more money for their

efforts, small arenas were built that featured rings demarcated by wooden barriers or heavy ropes. The current ring, with four (or occasionally three) ropes tied to turnbuckles on corner posts, is the descendant.

Although sanctioning bodies mandate the size of boxing rings, professional wrestling has no such requirement. In many venues, the same rings are used for boxing and wrestling. Amateur wrestling is done on mats laid across a floor. Ironically, the action in amateur wrestling is demarcated by a circle, yet it isn't called a ring.

None of this makes much sense without the historical perspective. That's probably why the most common slang term for the ring in professional wrestling is: "The squared circle."

Submitted by Virginia Graeber of Giants Pass, Oregon. Thanks also to Adam Rawls of Tyler, Texas; and R.J. Mamula of Hammond, Indiana.

DAVID FELDMAN

Why Does Iced Tea Get Cloudy?

You don't have to read tea leaves to get an answer to this Imponderable. According to the Tea Association of the USA, when tea is brewed with boiling water, tannins are leached out of the leaves and are released into the water. If the water is hot enough, most of the tannins, including theaflavin, the antioxidant that many believe helps lower LDL cholesterol, is dissolved. But when ice is added to the brewed tea, the caffeine and theaflavin form tiny particles that are harmless but unsightly.

Even if the murk doesn't affect the taste or healthfulness of the tea, many experts are armed with solutions to bring your tea back to clarity. In *On Food and Cooking,* food chemist Harold McGee recommends brewing the initial tea at room or refrigerator temperature over several hours. This technique extracts less caffeine and theaflavin than brewing in hot water, so the caffeine-theaflavin complexes don't form in sufficient quantities to become visible in the chilled tea.

If you don't have the time to cold-brew, just leaving the hot tea out in room temperature before refrigerating will keep the particles from forming. And if you live in an area with hard water, try using filtered water instead—solids dissolve more easily in it.

If guests are coming and you open the refrigerator, only to find the dreaded cloudy iced tea, there's an instant fix—add a little boiling water. Of course, if you do that, then the tea warms a bit. But if you throw some ice in, it might get murky again.

On second thought, maybe it's best to serve Snapple.

Submitted by Susan Thomas of New York, New York.

What's Medicinal About a Medicine Ball? And What Do They Stuff in Medicine Balls To Make Them So Heavy?

Until recently, medicine balls had a hopelessly old-school image. They were associated with punch-drunk boxers and obsolete physical therapy. Now they are trendy again among physical trainers. Instead of the old brown leather medicine balls, equipment makers are promoting balls in psychedelic colors that Jimi Hendrix would approve.

Trainers are hailing medicine balls as just the medicine for improving strength, flexibility, and reflex time in their athletes. Unlike most weight training apparatus, medicine balls encourage a full range of motion. While bodybuilders are often concerned with maximizing the size of muscles, trainers use medicine balls to promote functional training, resulting not only in better performance in sports, but in superior health, as multiple muscle groups are engaged with every use of the medicine ball. Every time you lift, throw, or catch a medicine ball, you have to stabilize the torso. Strengthening the core of back, abdomen, and hips seems to be the Holy Grail of fitness trainers these days, not only to promote strength, but to prevent back injuries that eventually plague most people as they age.

We corresponded with Cheryl L. Hyde, president and CEO of White Dolphin, Inc. and Academy Fitness, who loves not only the benefits of medicine balls, but their convenience and flexibility:

> Medicine balls and their "offspring" are very effective tools for increasing fitness levels. They can be used to increase strength, coordination, endurance, and even flexibility, depending on what you do with them.
>
> Newer is not always better and that is true of the medicine ball, which is useful in so many fitness and exercise settings. They take up less room and cost less than some types of equipment that are used for the same type of result. For instance, compare squats with a medicine ball versus squats with a bar and free weights or a squat rack. Now add a toss in the air as you stand up out of the squat and catch as you lower down and you get deltoid, triceps, biceps, trapezius, as well as some lats, pecs, forearm, and hand muscles included with the glutes, quads, and hamstrings. In addition, the core muscles get a workout due to the stabilization of the upper body. That is almost a total body workout. Pretty cool, huh?

We're not sure "cool" is the word we would have used, but we get the point. Hyde adds that unlike free weights and machines, medicine balls allow realistic motions and unrestricted movement. Trainees, for example, can attempt explosive and ballistic movements with the medicine ball.

HISTORY OF THE MEDICINE BALL

We weren't able to pinpoint the exact date when the term "medicine ball" was coined. As far back as 1000 b.c., Persian wrestlers trained with animal bladders stuffed with sand (we kid you not—as we discussed in *Why Do Dogs Have Wet Noses?*, early footballs were made from inflated cow bladders). The father of medicine, the ancient Greek Hippocrates, stuffed animal skins with sand and promoted the benefits of throwing and catching the "ball" (some think the Hippocrates connection is how the word "medicine" became attached to the ball). Not too long after, the Romans played games with balls called *paganica,* which were oblong medicine balls stuffed with feathers.

The credit for the invention of the modern medicine ball is given to a colorful figure, William "Iron Duke" Muldoon, a New York City police officer who was such an accomplished wrestler that he quit the force in 1875 to become the first person ever to be a full-time professional wrestler. Muldoon not only traveled with carnivals, taking on all comers, but eventually participated in the first wrestler versus boxer contest, ending his match with the most famous boxer of his time, bare-knuckles champion John L. Sullivan, when Muldoon quickly body slammed the hapless Sullivan.

Among his many other activities, Muldoon eventually became a boxing trainer, and developed several apparatus that live on to this day—the heavy bag and the medicine ball, made of heavy leather and stuffed with sand. Cheryl Hyde would probably not approve of

Muldoon's techniques. He not only had boxers throw and catch the medicine ball, but Muldoon dropped it on the boxers' abdomens, mimicking the explosive thrust of a punch.

Medicine balls soon caught on with a wider audience than boxers. YMCAs and schools prized the relative cheapness and portability of medicine balls. They were particularly popular on transatlantic ships during World War I, among both military and cruise ships. Perhaps nothing popularized them more than a presidential product placement. Shortly after his election, while aboard a ship returning from South America, president Herbert Hoover observed a game of "bull-in-the-ring," in which one player in the middle of a circle of competitors attempted to intercept a pass of the medicine ball.

Inspired by the game he saw and, perhaps, by his burgeoning waistline, Hoover rustled up members of his cabinet, Supreme Court justices, and other high officials, and invented the game of "Hoover-ball" in 1928. Teams of two or four competed every morning but Sundays, starting at 7:00 a.m. and stopping promptly at 7:30 a.m., regardless of the score. One player "served," throwing the six-pound medicine ball over a net on a tennis-sized court. The receivers had to catch the ball on the fly and return it immediately. The first team to allow the ball to drop inside the court lost a point.

Hoover credited his daily game for keeping his weight, which had ballooned up to 210 pounds at the beginning of his term, to a trim 185 during his administration. Hoover-ball became popular among even hoi polloi in the United States, but Americans seemed to tire of the game as they did of the President, who was not re-elected. Still, the Herbert Hoover Presidential Library Association sponsors a Hoover-ball championship every year in West Branch, Iowa.

OLD STUFFING

The engineering problem with medicine balls has always been how to stuff them to provide minimum fuss and maximum weight. In the good old days of inflatable bladders, the "balls" were probably filled with whatever was handy, most likely dirt or sand. A manufacturer of medicine balls, Healthtrek, claims that Hippocrates filled his bladders with sand. Sand continued to be popular filler for medicine balls, from the days of William Muldoon to the present. Other popular fillers included rags and kapok, the soft and fluffy fiber from the kapok tree, which is still a popular fill for upholstered furniture.

NEW STUFFING

Medicine balls come in a bewildering array of choices. Cheryl Hyde says that she likes gel-filled balls made by Thera-Band: "They are small—the eight-pound ball fits in the palm of my hand, and I don't have big hands." Others are the size of basketballs. Some have handles, most do not. A few still sport a leather exterior, while others have inches of rubber on the outside, and are hard enough to break a jaw or a rib if dropped at an inopportune time (or place).

Sand (along with dry air) still survives as a fill for some medicine balls, particularly ones not designed to bounce, but the state of the art today seems to be gel. Once medicine balls became a commercial product, especially for the health and fitness markets, an image of proper hygiene was essential, and customers became pickier about the composition of the fill. MediBall claims to have been the first gel-filled medicine ball—their smallest ball, five inches in diameter, weighs two pounds, and their fifteen-pound ball is nine inches.

No one asked us, but: What do Pampers and MediBalls have in common? According to MediBall,

> The balls are filled with an aqueous gel composed of potassium polyacrylate and water. It is non-toxic and non-hazardous, the same material is used as an absorbent in baby diapers. Should not hurt you unless you eat too many of them.

Hey! Leave the jokes to us.

Submitted by Taryn Losch of Gettysburg, Pennsylvania. Thanks also to Reena Mudhar of Germantown, Maryland.

What Is the White, Usually Oval Patch Found on the Bottom of McDonald's Hamburger Buns?

At first, our friends at McDonald's were a little reluctant to answer this question, as their buns are baked locally by independent suppliers. So we moped silently and ran to the trusty American Institute of Baking. Kirk O'Donnell, who is the only person we've ever encountered whose expertise includes strategic training, plant management, and hamburger buns, graciously replied:

> Hamburger buns often have a small white patch on the
> bottom due to the way they are processed. The dough is very
> soft, and when it is conveyed into the pans, it tends to entrap
> some air between the pan and the dough. The white patch is

simply the area where some air has kept part of the dough from being in contact with the hot pan in the oven.

Eventually, McDonald's did get back to us and confirmed O'Donnell's explanation.

We ran this Imponderable by Fraya Berg, food editor at *Parents* magazine. She mentioned that sometimes white patches on baked products can be caused by extra flour from the mold or cutter used to keep dough from sticking. Berg mentioned that in the *Parents* test kitchen, "We use a brush to get rid of excess flour when rolling cookies as we did tonight"—all to avoid the dreaded white patch.

Submitted by "Gooshie," via the Internet.

DAVID FELDMAN

Why Do Some Places, Such as Newfoundland, Australia, India, and Parts of the Arab World, Have Half-Time Zones? Why Do Some Large Countries Have Only One Time Zone?

As we write this, it is 6:40 p.m. in Pittsburgh, Pennsylvania, and three hours earlier, 3:40 p.m., in Reno, Nevada, about 3,800 kilometers west. While we're toiling away, the folks in Shanghai, China, are just waking up. It is 7:40 a.m. in Shanghai on the East Coast of China, but it also 7:40 a.m. in the city of Uramqi, located 3,800 kilometers to the west of Shanghai. The folks in Sydney, in eastern Australia, are already at work—it's 10:40 a.m., but Perth residents in the West are still commuting at 8:40 a.m. How about the folks in central Australia? For those in north-central Australia, such as residents of Darwin, it's

9:10 **a.m.** (Australian Central Standard Time). South of them, in Adelaide, it's 10:10 **a.m.** (Australian Central Daylight Savings Time).

If you are catching our drift, you might have already come to the conclusion that time zones are not uniformly applied throughout the world. When such large countries as China and India have only one time zone each, and places like Australia, India, Iran, and Newfoundland feature half-time zones, then something besides scientific considerations has affected the way folks tell time.

The notion of uniform timekeeping throughout the world is a surprisingly recent phenomenon—until the late nineteenth century, towns would set their own standards. If there was a big clock in the central square, an official would try to calibrate noon to when the sun was directly overhead, and hope to be reasonably accurate.

No one seemed to be highly perturbed by this haphazard arrangement until railroads demanded a more precise way of scheduling routes, especially in the United States and Canada. It was a mite difficult to print train schedules when a ten-minute trip could send passengers to their destination earlier than they had left! Ruth Shirey, of the National Council for Geographic Education, wrote *Imponderables:*

> Before railroads allowed us to travel fast enough that time
> zones were standardized so that trains could be scheduled,
> communities all over the world went by local sun time. In
> many parts of the world, people still use local time.

The inventor of our time zone system was a railroad engineer, Canadian Sandford Fleming. His notion was elegant in its simplicity: If the earth takes twenty-four hours to rotate, and there are 360 degrees of longitude, why not create twenty-four time zones of 15 degrees of longitude each? Sure, the zones in the extreme north and south would be infinitesimal, but only a few scientists, polar bears, and penguins might complain.

In 1884, President Chester Arthur convened a Prime Meridian Conference in Washington D.C. to standardize the concept Fleming had developed just six years before. And although not quite unanimous (San Domingo voted against, and Brazil and France abstained), twenty-two other nations voted for naming Greenwich, England as the location of the Prime Meridian.

But there has never been total compliance with Fleming's scheme. China should have five time zones but it has one (in the western part of China, the sun is often overhead at 3:00 p.m.). India is the second largest country to have only one time zone (in Fleming's scheme, it should have two), and that one is a half-time zone.

According to Shirey, by far the most common reason for half-time zones (or "offset time zones") is to shift key cities closer to their actual sun time. For example, all of Newfoundland is one-half hour ahead of Atlantic Standard Time, the zone used by the other Maritime provinces in Canada. Newfoundland lies on the eastern edge of its geographically correct time zone, so it chose an offset time zone to better reflect its actual sun time. Newfoundland's offset time zone is so popular that when the government tried to change to Atlantic Standard Time to conform to Labrador and the other Maritime Provinces, the public outcry prevented it from happening.

The offset time zone in central Australia has a different story. Thomas H. Rich, a curator at the Museum of Victoria in Melbourne, Australia, took an interest in this Imponderable, and unearthed proceedings from the South Australian Parliament in 1898, that showcased the debate in Adelaide about whether to reject the Fleming system that it had adopted in 1894. The reason for the opposition in Adelaide had nothing to do with the sun and much to do with dollars:

> commercial men who received cable advices from Great Britain were put to great disadvantage under the present system as compared with business men in the other colonies . . . commercial cablegrams are generally delivered in the morning, and in consequence of the present arbitrary law by which the Adelaide time is made one hour later than that of Melbourne and Sydney, South Australian merchants are placed at a great disadvantage, their competitors having one hour to act on the cablegrams before the local commercial men are in receipt of theirs . . .

The original reason for the offset is long past, but reversals of offset time zones are rare—all politics is local.

Meanwhile, anomalies exist all over the globe. Nepal, just to the west of Bangladesh, has a fifteen minute offset. Russia's entire country is one hour off, sort of—it is on permanent daylight savings time. But our favorite brouhaha is the controversy about Daylight Savings Time in the United States. Until 2005, Indiana had a byzantine structure, in which most of the counties in the Eastern time zone refused to switch over to DST, but there were renegades who did, along with a few "traitors" from the Central time zone as well. Indiana's localities gave schedule makers a nightmare.

Arizona is the last of the original forty-eight states not to endorse Daylight Savings Time, but even here, there is a holdout. The Navajo Nation in Arizona observes Daylight Savings Time, but the Hopi Nation, which lives within the Navajo Reservation, does not. If we can conclude anything about this topic, it is that although astronomy might have inspired our attempts at measurement, in practice politics, business considerations, and human psychology dictates how we tell time.

Submitted by a caller to the Larry Mantle Show, KPCC-FM in Pasadena, California. Thanks also to Jim Sears of Belton, Missouri; Anthony Bialy of Kenmore, New York; Joe Koch of Ellington, Connecticut; Nicole Nims of Culver City, California; Peter Darga of Sterling Heights, Michigan; Claxton Graham of Mt. Holly, North Carolina; and John Buchanan of Hamilton Square, New Jersey.

How Did the Candy Snickers Get Its Name?

Snickers, introduced in 1930, was named after a horse, albeit a well-connected one. Snickers was the favorite horse of a family named Mars, who just happened to own a certain candy company.

Submitted by Heidi Zimmerman of San Diego, California. Thanks also to Justin Tedaldi of North Massapequa, New York; Alexa Steed of Colorado Springs, Colorado; and David Weinberger of Los Angeles, California.

Why Do Motorcyclists, Especially Those Riding Harley-Davidsons, Rev Their Engines While Waiting at Stoplights?

We were excited when we received this Imponderable from one of our most prolific correspondents. We've always wondered about this, too, so we used the power of the Internet to connect with hundreds of motorcyclists directly at the forums of Motorcycle-USA.com, GSResources.com (for owners of Suzuki's GS Series), and Harley-Zone.com, for Harley-Davidson owners and fans.

The responses were all over the map and revealed a lot not only about motorcycles, but about the crosscurrents among motorcycle enthusiasts—here is a subculture with many subcultures within it. In most cases, we are not quoting our sources by name, often for obvious reasons.

TECHNICAL REASONS FOR REVVING

The first obvious question is: Are there any mechanical or technical reasons for revving a motorcycle when it has come to a complete stop? Yes, say many of the respondents, and argue that this is the only reason why *they* do so.

> If my carbureted engine isn't warmed up completely, sometimes I need to rev to prevent stalling at the first stop sign.

> I do it to get my RPMs up in anticipation of a green light.

> The rev set originated because the bikes would come in one of two flavors. There were the street racer set (which generally rode British motorcycles) and the Harley-Davidson crowd.
>
> The former would run "hot" cams, which would give them very good power at the top of the rpm range, but would make them stumble at idle. When the rpm's drop below a certain level, the oil isn't flowing and engine damage is likely to result.
>
> The Harley-Davidson set have V-twin engines that utilize a single crankpin design (generally). This causes the bike to "lope" heavily. The older naturally aspirated units . . . were prone to stumble and stall at idle. . . . Due to the engine's tendency to stall at very low idle speeds . . . some throttle maintenance was required to keep the bike running.

Now, most modern Harley-Davidsons are fuel-injected, and tend not to stall out when idling, but most of the bikes with earlier technology are still on the road. Several Harley owners admitted that now that their bikes are no longer stalling when stopped, they've lost their "excuse" to rev.

A few other technical reasons for revving, nicknamed "blipping" by riders, concerned auditory matters.

> I've found myself on my bike with my helmet on and it
> idles so low that I can't tell if it's still running, so I rev it to
> make sure.

And sometimes a blip is a "gentle" reminder to others:

> If someone in front of me isn't moving on a green light, I'll
> rev my engine before honking.

AESTHETIC REASONS FOR REVVING

Motorcycle riders might have reputations as macho tough guys, but their responses read more like opera criticism. Many of them, especially Harley owners, were obsessed with the sound of their bikes:

> When I was eight or nine, I was nearly asleep in the backseat
> of the family station wagon, when a group of motorcycles
> went by and grabbed my attention.
> I can't tell you what color the bikes were or recognize any
> of the faces but I will never forget the "Sound." My Dad told
> me they were Harleys and he knew because of the unmistak-
> able Sound. That Sound only comes from a V-twin (air-
> cooled) engine.
> Now, thirty-five years later, I still get that same feeling
> when I hear "It." I think that most of us who ride Harleys
> love the Sound and don't hesitate to share the Sound. Kinda

like flexing your muscles at the beach, I guess. People in cars often encourage it by giving the universal signal, the throttle twist. And when they do it's never just one twist. It's always two or three. I don't know about intersections, but if you pass me under an overpass, you will likely hear a couple of extra revs with the clutch in. Gotta love that Sound.

Harley Dave is a sharing kind of guy:

> I rev it up and when my wife is on she says, "Why do you do that?" I say, "Because I love that sound!" When I am driving and some kid waves and grins because he (or she) is thinking "Someday," I do it for them, too! Everyone loves the sweet sound of "Screamin' Eagles" on a calm, quiet, morning. If they don't, they better get used to it in Harley Dave's neighborhood.

Riders were so passionate about the Sound (which they seem to capitalize as if it were a proper name) that we asked them to be honest and say whether they would still rev even if they found themselves stopped at a traffic light on a deserted road. We received many responses like this:

> It's just the sound that I like. I don't care if there's a hundred people around or it's just me, the sound makes my blood flow.

One enthusiast brushes aside the mechanical arguments:

> People can say their bike runs rough, it does not idle correctly, or a million other excuses. They are all lying. The truth is they do it because they love the way it sounds.

REVVING'S ALL ABOUT ME

Why ride a motorcycle at all, let alone rev it, if you don't like the sound? But some riders are willing to admit the sound is sweeter when you draw attention from others:

> [There is] Nothing cooler than sitting at a light at Third Avenue and Fourteenth Street when all those people walk by and look at you on your scoot. You give them a quick rev as acknowledgment or to catch the attention of a good-looking young lady.
> What's even funny to me as I sit at a light and a crotch rocket pulls up with mosquito-sounding pipes. They take off like a bat out of hell and you gently let off the clutch and everyone watches you coast in to the sunset. Now, *that's* cool.

And it doesn't hurt if there's a cute member of the opposite sex to admire your coolness (even if it's your spouse):

> Love that thunder!!! That's half the greatness of riding a Harley . . . The Wife just about gets *too* excited, especially when I start the Deuce [a particularly stylish Harley Model]. If I'm going with or without her, she is always in the garage when the bike comes to life. Sometimes between the thunder and the leather, I'm lucky enough to get out of the garage.

But not all riders view blipping positively, and see it as a sign of insecurity rather than confidence:

> It's an extension of their manhood. Just like the people who go out and put huge tires and lift kits on 4×4's and never take them off the road. They're making up for a shortage in their anatomy.

One Harley lover is worried that attention-getting behavior by others ends up stereotyping riders:

> In the old days of weak ignition systems and poor carburetion, bikes had to be revved to keep them from stalling. Those days are are long past but a lot of wankers out there still rejoice in what most real riders refer to as stoplight motorcycle monkey spankin', dolphin waxing (choose your euphemism). There are riders out there that dropped $20,000 on their bike and another $20,000 on chrome and mods [often modifications to make the bike nosier] and still suffer the insecurity and fear that they may not be noticed. The straight pipe stop sign throttle rapper is more than likely the guy who routinely gets dumped on at work, nagged by his old lady and ignored by his kids saying, "please look at me, I'm here. . . . I'm going all Sigmund Freud on you.
>
> Just as likely, this behavior is another by-product of our ready-made subcultures that we no longer invent ourselves but rely on marketing and media to create and spoon-feed us. [The] motorcycles, politics or music scene we get—most of it is prepackaged. The resultant cultural behavior is predictable.

REVVING TO PUNISH

Motorcyclists have a term for automobiles—"cages." More than a few folks on the forums have contempt for the folks inside the cages ("cagers"), if not for the cages themselves.

Several riders were willing to admit that they used the blip to punish annoying cagers. The single biggest offender seemed to be cell phone users:

The amount of throttle, duration and frequency of my "spanking it" at the light is proportional to the idiots in their cages on their cell phones. It's so funny to see the windows go up [and] the phone shift to the other ear. Did I tell you I hate cell phones in vehicles?

But there was plenty of hate left for aggressive drivers:

If I am at a stoplight behind a car with a body kit, tacky paint job, and a bunch of "sponsorship" stickers, I will often rev the engine to get the driver's attention. I have found that in 100 percent of such cases, the result is the moron in the car will take off as fast as possible, trying to impress me. This allows me to take off a pace closer to my normal (brisk) one. Also good for getting rid of the same type of driver when they are next to you and you want to get in that lane to turn.

and inattentive drivers:

I don't blip at lights to enjoy the sound or to impress anyone. I do it to get the attention of the blond cager, with the cell in one ear and one eye looking at her makeup in the rearview, and the other eye on the light as it changes. I do it to remind her there is a motorcycle there so hopefully she won't run me over until the SUV in front of me gets out of my way. Also to wake up the old cager in the SUV after he fell asleep at the light.

or conventional people:

I do it to annoy soccer moms [this was followed by a smiley face emoticon].

or even *Imponderables* book authors:

> Look around next time, there's probably some nob in a
> BMW on a cell phone drinking a latte I'm trying to piss off,
> or maybe it was you. Get back in your 4-door shopping cart
> dude, or better yet buy a bike and join the dark side . . .

Gee, we wish we had a BMW. And if we bought a bike, with our coolness quotient, it would probably be a Schwinn with training wheels.

REVVING FOR FREEDOM

Scan the contents of motorcycle forums and the word "free-
dom" is omnipresent. Many riders insist that they blip not to annoy
others, but to luxuriate in their favorite F-word:

> We spend thousands to customize our bikes just the way
> we like them to please *us!* This includes the sound of the mo-
> tor. Yes, I give it a little blip at lights from time to time, be-
> cause I paid for that melodic sound with my hard-earned
> dollars, not as some have suggested that it somehow makes me
> feel like a big man—my lady does that for me, not my bike.
> Riding has always been an extension of *freedom* to me, not
> my manhood.

REVVING IS PART OF THE RIDING EXPERIENCE

Many riders don't separate revving from the experience of being on the open road: Psychotherapist Brenda L. Bates, who rides
herself and specializes in treating bikers who have encountered accidents, wrote to *Imponderables* via e-mail:

Many of us are what is called "motorheads." We love the feel of the engine and take any opportunity to manifest the sensation that revving those rpm's gives us. Many car drivers don't care or know much about the engine in their auto. But motorcyclists can be different in that we truly appreciate the sound and feel of a good engine.

We thought blipping must be a way of revving up the metabolism of the rider as well as the bike, but competition racer Kris Becker (http://www.krisbecker.com) told us that for most riders, it's exactly the opposite:

> Harley riders (and any motorcycle rider, any brand), whenever stopped at lights or in a field when off road, have just come from an intense experience. Now they are "idling" mentally and mechanically, they rev the engine to fill the hole in their sensory input left after the movement and sensory intensity of the riding experience. Note that as soon as they are on a highway, at 50, 60, 70 miles per hour, they don't look around much, they are relaxed, they have found the "zone." Watch them stop and again, a little rev, some foot tapping, looks around, mouth popping, and "jonesing" for the reestablishment of the road experience.

REVVING IS—I HAVE NO IDEA!

Next to those who professed to love the Sound, the most popular answer was: "I don't know," but the love of blipping is the expression of an inchoate joy:

> We do it because we can, honestly. I sometimes do it in my toy car for the sake of it, but on the motorcycle, it's one of

those, "Hey, I'm here" things. It's our way to express ourselves while stationary.

I do it. I'm a blipper. I do it for lots of reasons. To be cool, to let others know I'm there and for well, I don't know. I just do it. I guess that makes me an addict. . . . the first step in recovery is to admit . . . I'm a blipping blipper . . .

THE CASE AGAINST BLIPPING

This Imponderable exposed divisions within all three online communities of bikers. Cyclists are acutely aware of their image among nonriders, and some commented on this schism:

As you may gather from some of the . . . posts, many riders are quite defensive about this—there is an attitude that they have the right to make noise, and damn any objections. In the motorcycle world, aftermarket mufflers and exhaust systems are commonplace—a lot of guys will install them strictly for loudness, with no concern about performance gains. . . . And of course, once you've spent hundreds or thousands of dollars on a loud exhaust system, you want to hear it—so you'll sit and rev it at red lights, getting envious looks from any guy within sight and dreamy eyes from the women, as you sneer from beneath your do-rag—but oops, I digress. As I said, some of us ride motorcycles because they're fun to ride, and don't feel that annoying the non-riding public with excess noise is required, or politic.

Most of our motorcycle enthusiasts were in the pro-blip camp or the anti-blip minority. Few were sympathetic to both sides. We exchanged a fascinating series of e-mails with "Hawkster," who is attracted to the scream and the danger of motorcycles and other macho symbols, but is also aware of their pitfalls:

> As a former range safety officer and instructor at an outdoor shooting range and as a current riding instructor, I've noticed a lot of similarities in the mind-sets of both shooters and motorcyclists. I know that if it's both noisy and dangerous, it will probably draw my interest!
>
> Many people own either guns or motorcycles because they feel empowered by them. Whether or not this is a good thing is a matter of some debate among instructors. I believe that it ultimately comes back to how a particular object or pursuit makes us feel—I've seen the "I'm a bad-ass" look on many a new shooter's face. Unfortunately, empowerment only comes through proficiency, but most don't seem to realize that.
>
> Many riders, particularly the newer ones, are under the illusion that they are able to buy into the movie- and television-created "biker mystique" with the purchase of a machine. I call this the "bad-ass by proxy" syndrome . . . Crash statistics indicate that it doesn't usually bode well for their riding career.
>
> Much of my time at the shooting range was spent stomping out myths about guns. Oddly, I seem to spend just as much time doing the same during motorcycle classes.

We were impressed with how thoughtful Hawkster was, so we asked him why he's so attracted to the noise and danger himself. He replied:

> As an instructor, I can get all the attention without it becoming socially unacceptable (Are revving bikers straddling

that line?). The need for attention is probably a "shadow motive" for many actors, activists, and teachers.

Regarding shooting and riding motorcycles on the street and in competition: I think that some people just function best in a threat-rich environment. I'm that type of person. Many of us have been labeled as ADD (Attention Deficit Disorder). I think it really gets your attention when your life is on the line. We enjoy that. I consider riding to be transpotainment.

I don't think that anyone ever really grows up. We're just forced into acting like it by cultural mores. Riding a motorcycle permits us to be (borderline) socially acceptable kids.

The Sound is both the boon and the bane to Harley devotees. A few years ago, when Harley-Davidson celebrated its one-hundredth anniversary, upstart motorcycle manufacturer BMW posted this billboard on the Interstate: "We'd say congratulations on your 100th. But you wouldn't hear us."

Submitted by Douglas Watkins, Jr. of Hayward, California. Special thanks to all the participants of the Motorcycle USA, GSResources.com, and HarleyZone message boards.

Why Are Salmon Pink?

Salmon are pink for the same reason that lobsters, crabs, crawfish, shrimp, and for that matter, flamingos are pink: They are what they eat. Salmon feast on shrimp, krill, and other small fish that are full of asataxanthin, a carotenoid (much like the beta carotene that give carrots their orange hue) that not only adds the reddish pigment to the salmon's flesh, but provides plenty of vitamin A.

Skeptical *Imponderables* readers might think this answer is fishy for one big reason. Most salmon we see in supermarkets, fish stores, and restaurants are not wild salmon, but farm-raised. These farm-raised salmon are not pampered with shrimp cocktail feasts—they are given a less expensive commercial feed.

That feed is infused with plenty of asataxanthin for the most compelling of reasons: the color of salmon is a major factor in its consumer appeal. In a paper, "Salmon Color and the Consumer," Stewart Anderson of Hoffman-La Roche Limited wrote:

> Consumers perceive that redder salmon is equated to these characteristics: fresher, better flavor, higher quality, and higher price.

Anderson makes clear that redder salmon doesn't taste any better and is not any fresher, but the bottom line is clear, which is why you are unlikely to see any pale salmon, from whatever source, in your store or on your plate.

Submitted by Simon Parker-Shames of Ashland, Oregon.

Why Do Parrots and Other Birds Mimic Human Speech and Other Sounds?

Wouldn't it be cool to be lost in the jungles of Belize and have a yellow-headed parrot assist us with a timely: "Hey buddy, bear left!"?

Cool? Yes. Likely? No way. As far as we know, birds, even chatty ones like parrots and mynahs, do not mimic human speech in the wild, but they do imitate other sounds, and especially other birds.

In the wild, birds mimic for a variety of reasons. Depending upon the circumstances, birds feigning the songs of other species can attract members of the opposite sex, fool predators into misidentifying the species of the singer, or ward off competitors for food or territory. Since humans are usually capable of differentiating between the song of a mimic from the song of the bird the mimic is imitating, it's likely that the impersonation isn't fooling other birds (whose hearing is much more sensitive than ours) either. But like humans, birds would rather avoid confrontation or competition than face it head on—mimicking clearly works.

Birds are flock creatures. Closeness within its species is crucial to a parrot's survival in the world: While one parrot focuses on foraging, others look for predators. When taken out of the wild and into a household, birds will imprint with their human owners. Mimicking seems to be a way to foster closeness with their human "flockmates." As Todd Lee Etzel, an officer of The Society of Parrot Breeders and Exhibitors, wrote *Imponderables*:

> The pet bird becomes imprinted with human vocalizations
> and hence mimicking takes place due to the bird's desire to be
> part of the "flock," even though it is not a natural one. Keep
> in mind that most species of parrots are highly social and the
> need for social interaction is so strong that innate behavior is
> modified to fit the situation.

If pet birds are motivated by social interaction, why do they tend to mimic at least as much when their owners leave the room? Animal behaviorist W. H. Thorpe offers a theory:

> as they develop a social attachment to their human keepers,
> they learn that vocalizations on their part tend to retain and
> increase the attention they get, and as a result vocal produc-
> tion, and particularly vocal imitation, is quickly rewarded by
> social contact. This seems an obvious explanation of the fact
> that a parrot when learning will tend to talk more when its

owner is out of the room or just after he has gone out—as if he is attempting, by his talking, to bring him back.

Thorpe's theory advances the notion that bird psychology differs little from infant psychology, which isn't farfetched in the least. Dr. Irene Pepperberg, a research scientist at MIT and a professor at Brandeis University, has studied and written about the abilities of African Gray parrots, particularly her oldest bird, Alex. Just as other scientists have proven that primates are capable of complex communication, so Pepperberg has shown that parrots can do far more than mimic. If shown two blocks that are identical except for their hue, and is asked what is different about them, Alex will answer "Color." He has mastered numbers, and shapes, and locations, and the names of objects, and has learned to ask, or more accurately, demand them: "If he says that he wants a grape and you give him a banana, you are going to end up wearing the banana." Pepperberg estimates that Alex's cognitive ability is comparable to a four- to six-year-old child, with the emotional maturity of a two-year-old.

Pepperberg plays a version of the classic shell game with Alex and other parrots, hiding nuts below one of three cups. Alex usually succeeds at spotting the nut, except when the experimenter doesn't play fair. Sometimes, the scientists will deceive the parrot by sneaking the nut under another cup while distracting him from the subterfuge:

> So Alex goes over to where he expects the item to be, picks up the cup, and finds that the nut is not there; he starts banging his beak on the table and throwing the cups around. Such behavior shows that Alex knew that the object was supposed

to be there, that it's not, and he's giving very clear evidence that he perceived something, and that his awareness and his expectations were violated.

In the wild, parrots use their wits to evade predators and find food and mates. The need to solve problems might be as innate in a parrot as its mimicking skills. If a parrot's cognitive skills are as great as Alex has exhibited, Pepperberg implores pet owners to provide the proper stimulation:

> I try to convince them that you can't just lock it in a cage for eight hours a day without any kind of interaction. I don't mean just interpersonal interaction, or having other birds around; parrots have to be intellectually challenged.

Etzel thinks that parrots and other birds mimic our language and other sounds in their environment "simply because they are able to do so. It might even be a form of entertainment for them." Indeed, parrots might be musing about us while we are "training them":

> Doesn't my owner look silly constantly repeating "Polly wanna cracker?" Oh well, might as well go through the drill if it's going to end up with some tasty carbs down my gullet.

Submitted by Wayne Lipe of Long Beach, New York.

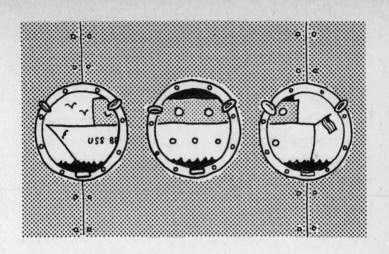

Why Are Portholes Round?

W indows have two main jobs: to let in light and fresh air. This doesn't seem too much to ask of a sheet of glass surrounded by a base frame, but ships pose a few extra little problems. Obviously, portholes on ships need to contend with water (windows that don't seal properly aren't too popular on ships, especially windows that are underwater at times). But a more pernicious danger is the movement of the boat itself.

When most ships were made out of wood, portholes were usually rectangular. But once steel construction came into vogue, sharp corners morphed into arcs. Wood may not be as hard as steel, but it has one feature that makes it more porthole-friendly: Wood absorbs the stresses of the rocking of boats on the sea far better than metal. When steel hulls came into vogue in the late nineteenth century, sailors discovered quickly that stress fractures were endemic to rectangular portholes, starting at the corners. Round portholes, on the other hand, distribute the stress evenly, and naval architects figured out the spherical solution quickly.

Rectangular portholes are far from extinct, however. Some wooden ships still have them. And cruise liners often sport large rectangular windows on decks. But the more violent the weather a boat encounters, the rougher the seas it navigates, and the farther down in the boat it resides, the more likely the porthole is to be cornerless.

Submitted by Mike Roberti of Duarte, California. Thanks also to "Carol," via the Internet.

Fly
Grandma Class!
• window curtains
• fluffy pillow
• antimacassars
• lots of loving
  encouragement
  for your business
  projects, too!

Now We Know Why Ships' Portholes Are Round. Why Are Airplane Windows in the Passenger Cabins Oval?

May we exert executive privilege and pose this Imponderable, which occurred to us after researching the last Imponderable? Surely, airplane windows are subject to extreme pressures in the air. Ken Giesbers, our mole at Boeing, confirmed it:

> Rounded holes in the thin fuselage are structurally more sound, and much less prone to stress fractures. Stress fractures in a pressurized cabin can lead to explosive decompression and outright structural failure.

The seriousness of the issue was highlighted when the first commercial jet, the De Havilland Comet of Britain, was plagued with three crashes shortly after its introduction in 1952. Much to their shock, thorough investigations revealed that the main culprit in all three crashes was likely metal fatigue. And most of the deterioration started at the corners of the Comet's large, rectangular windows. The Attorney General's report concluded that

> up to 70% of the aircraft's ultimate stress under pressure was concentrated on the corners of the aircraft's window.

The Comet was then redesigned with a stronger fuselage and round windows.

If round windows are best, why do Boeing and Airbus provide us with oval ones? According to Giesbers, they just aim to please:

> Having round windows would necessarily mean more solid material in the gap between windows. By elongating the windows vertically, aircraft designers can provide more viewing area (more surface area devoted to windows) and also better accommodate passengers of differing heights.
>
> Boeing is very proud of the large (19″ × 10.3″) windows its new 787 will have. The windows can be larger because the 787 will use more composite materials than before. Boeing makes no bones about the reason [for the big windows]: a better experience for the passengers.

Submitted by Dave Feldman, of Imponderables Central.

Why Don't Trains Have Cabooses Anymore?

All but the youngest of *Imponderables* readers will remember the little caboose, the last car on every train. Such is the appeal of that humble car that most model sets, even of contemporary trains, still feature cabooses, even though they have mostly disappeared since the early 1990s. In their stead, the back of most trains feature a skeletal open car with a flashing light.

What purpose did cabooses serve in the first place? Up until the mid-1980s, the typical freight train used to have at least four crew members. An engineer and a brakeman sat at the front in the locomotive; the other brakeman and a conductor brought up the rear in the caboose. Before the advent of radio communications, the two men in the caboose were eyes and ears for the engineer, and vital

communication was handled via hand and lantern signals. Many cabooses featured cupolas, which served as observatories for the crew,
who were on the alert for any sign of smoke, fire, or dragging equipment. The caboose also housed much of the train's valuable technology, including an emergency brake, gauges to measure brake
pressure, and the tools needed to make repairs to the train.

The caboose also served as a combination bedroom, office,
kitchen, and bathroom as well. Many were equipped with stoves for
cooking, a desk for paperwork, a latrine, and bunks for the brakemen and conductor.

Radio communication and, later, sophisticated telemetry devices eventually rendered the caboose obsolete. Perhaps the most important development in the demise of the caboose was the invention
of automatic rear-end devices (usually called "FREDs" (Flashing
Rear End Devices)). FREDs allowed the engineer to determine the
air brake pressure from gauges on his board in front, and often to
apply brakes in the back of the train—duties heretofore performed
by the brakeman in the caboose. These FREDs flash the red light
you see in lieu of the caboose on most trains today.

Another technological breakthrough that replaced human expertise is the Hot Box Detector (HBD), which automatically checks
for overheated wheel bearings or "hot boxes." If there is a problem,
the exact axle location is automatically signaled to the engineer.

Before sophisticated telemetry, a constant challenge was ironing out the slack on long trains. In the "old days," the conductor in
back would check for slack, but now the End of Train Device
(ETD) lets the engineer in front know when the back is moving—a
message display on the engineer's board lets him know.

Most states had laws mandating cabooses on trains well into the
1980s, but these slowly fell by the wayside as the big railroads convinced the Federal Railroad Administration that brakemen, and thus

DAVID FELDMAN

cabooses, were no longer needed. The motive of the railroads, of course, was financial. As Jeff Moore, webmaster of the High Desert Rails Web site (http://www.trainweb.org/highdesertrails/) put it:

> The elimination of cabooses saved railroads vast amounts of money. Supplying and maintaining cabooses cost a lot of money, as did having to switch them on and off trains and then storing and further switching them at terminals. Eliminating cabooses also meant that much less dead weight that the locomotives had to drive.

Alabama locomotive engineer Jerry DeBene told *Imponderables* that it costs about $2,500 a month to maintain a caboose: "Times that by a fleet of them and you can see why they were replaced." Some estimates for the cost of cabooses run much higher, up to six figures a year.

Although they are endangered species, neither the brakeman nor the caboose is totally extinct. On some lines, brakemen have been renamed the more generic, "trainmen." Charlie Tomlin notes that on the Burlington Northern and Santa Fe's Chicago division, the collector positions on commuter trains are the responsibility of brakemen and those brakemen are responsible for much of the switching: "Believe me, from having worked all of those jobs as a brakeman, there is plenty of work to do for two."

And Jeff Moore explains the main reason why some cabooses live on amidst today's technology:

> There are still a few applications where you will find cabooses in use today. These are primarily in situations where a train crew has to make a long reverse movement. Regulations require a train crew member to protect the movement by

riding on the last car, which can be dangerous under any circumstances. In such situations, railroads will generally provide a caboose, although more often than not the caboose has
been stripped of all hardware and crew amenities and is classified as a "shoving platform."

Submitted By Douglas Watkins, Jr. of Hayward, California.

Do Birds Sweat?

Nope. Not even when they are nervous.

Birds don't have sweat glands, so they can't sweat. But they have plenty of methods to cool themselves off. Birds are warm-blooded, like we are, and their normal body temperature is actually a little higher than ours.

Although you may have never heard it, birds also pant, just like dogs, and can cool themselves off in this way. And when birds fluff up their feathers, it isn't just to show off—fluffing allows air close to the skin so that even more evaporation occurs. These are the two most common ways that birds eliminate excess heat.

Birds are so active, and burn off so many calories while they fly about looking for food (including migrations that, for some birds, can require thousands of miles of flying), that it is a constant struggle for them to maintain the proper temperature. Hypothermia is a serious danger, so some species have developed specialized mechanisms to regulate their body temperature.

Have you ever seen the fleshy part of a bird's bill vibrate? Herons do this most visibly, but many other species, such as boobies and roadrunners, regulate their temperatures by this "gular fluttering." By vibrating the hyoid muscles and bones in their throat, gular fluttering achieves the same cooling effect as panting.

In *Why Don't Cats Like To Swim?*, we discussed how penguins' feet can withstand frigid conditions in Antarctica. But many other birds in warm climates use their feet to cool off. Martha Fischer, of Cornell Lab of Ornithology, explains:

> Herons and gulls can also lose a large percentage of heat
> through their feet. The veins and arteries in birds' legs and feet
> are intertwined and the blood flowing out to the extremities in
> arteries is cooled by blood flowing back to the body in veins.
> This is called countercurrent exchange (and is the reason ducks
> can stand on the ice without freezing their feet).

Fischer adds that birds are believed to have evolved from reptiles, which also do not sweat:

> In their evolution to their present state, selection has fa-
> vored physiological and morphological changes that enhance
> light-weightedness.

Dinosaurs would tend to agree.

Submitted by Lorelei Truchon, of Fairfax, California.

DAVID FELDMAN

Why Is the Earth's Core Still Hot?

We knew that the Earth's core is hot, but not how hot. Actually, even geologists don't know exactly what the temperature is, either, but they know it is almost as hot as the Sun. That's hot.

Luckily for us, the surface of the Earth is considerably cooler. After all, the surface is losing heat by being in contact with cooler air. But the core is continually heated by the decay of radioactive elements. Scientists believe that the original heat from the formation of the Earth is still being played out in these transformations. The Sun is likely to expand and burn us before the core of our planet cools off.

The rule of thumb has traditionally been that for every sixty feet in depth, the temperature of the Earth increases by one degree Fahrenheit. But this old approximation was based on distances that scientists could measure by drillholes. If this ratio held true all the way to the core of the Earth, its temperature would be 180,000 degrees Fahrenheit. Most geologists agree that the actual temperature is closer to a still unfrosty 9,000 degrees Fahrenheit.

Why are we more worried about future global warming when our own core is so incendiary? The amount of heat that reaches the Earth's surface is not enough to affect it drastically, and what heat there is quickly radiates to outer space. In their book, *Physical Geology*, Brian J. Skinner and Stephen C. Porter emphasize that the heat transfer is not uniform:

> The heat loss is not constant everywhere. Just as the geo-
> thermal gradient varies from place to place, so does the heat
> flow, which is greatest near young volcanoes and active hot
> springs, and least where the crust is oldest and least active.

Of course, the Earth's crust helps insulate the core from losing more heat. These cracks in the crust, like volcanoes and springs, are the few venues that allow us to glimpse the heat trapped in the sizzling core below.

Over the past few years, scientists have developed the ability to measure radioactivity more precisely through measuring particles called "antineutrinos." Scientists have long suspected that radioactivity might be responsible for all the heat generated at the Earth's core, and geologists are optimistic that eventually they will be able to map exactly where the energy is being generated. Even more recently, two University of California, Berkeley, scientists discovered that potassium can form an alloy with the iron in the Earth's core. Although Dr. Kanani Lee found that potassium might comprise only one-tenth of one percent of the Earth's core, "it can be enough to provide one-fifth of the heat given off by the Earth."

The University of California, Berkeley announcement of the potassium discovery presents the current consensus:

> The Earth is thought to have formed from the collision of many rocky asteroids, perhaps hundreds of kilometers in diameter, in the early solar system. As the proto-Earth gradually bulked up, continuing asteroid collisions and gravitational collapse kept the planet molten. Heavier elements—in particular iron—would have sunk to the core in 10 to 100 million years' time, carrying with it other elements that bind to iron.
>
> Gradually however, the Earth would have cooled off and become a dead rocky globe with a cold iron ball at the core if not for the continued release of heat by the decay of radioactive elements like potassium-40, uranium-238 and thorium-232, which have half-lives of 1.25 billion, 4 billion, and 14 billion years, respectively. About one in every thousand potassium atoms is radioactive.

As you have seen, scientists tend to take the long view. Although San Francisco biophysicist Joe Doyle later reiterated the "radioactive theory," his first response to why the center of the Earth is still hot was:

> There hasn't been enough time for it to cool yet.

Submitted by Lance Burpee of Lisbon Falls, Maine. Thanks also to Peter Gosling of Ann Arbor, Michigan; and Charles Scourfield of Moss Vale, New South Wales, Australia.

Why Do More Men Snore Than Women?

"Laugh and the world laughs with you. Snore and you snore alone."

No doubt about it. The delicate issue of snoring has caused fussin' and feudin' in more than a few relationships. Abigail "Dear Abby" Van Buren published a letter from "Frantic in Fresno," a woman who was married for fifteen years to a man whose snoring woke her up "wondering who was mowing the lawn," and who complained about having nightmares in which "a tugboat was stuck in the bedroom." The husband wouldn't allow Frantic to sleep in another bedroom, and wouldn't seek medical help for his condition, which, come to think of it, didn't keep *him* up nights.

Abby asked her readers for their comments, and received more than 5,000 replies. More than 90 percent of the respondents reported that they slept apart from their snoring mates. Abby concluded that "love is blind, but not deaf," adding that when women were confronted about snoring, they took it as an insult. In fact, their mates reported that most women wouldn't admit to snoring.

It's hard not to joke about snoring, but consider this: The loudest snore ever measured, 92 decibels, exceeds the United States' Workman's Compensation threshold for requiring ear protection on a worksite (90 decibels). By comparison, normal speech is approximately 40 decibels; normal city traffic, 65 decibels; a jet plane, 110 decibels; a jackhammer, 120 decibels; and the pain threshold (think of a heavy-metal rock concert), 140 decibels.

About one-half of adults snore occasionally, and about one-quarter regularly. The two most popular stereotypes about snoring happen to be true: Older folks, as we reported in *Do Elephants Jump?*, snore more than younger ones; and more men *do* snore than women. Almost twice as many men as women snore, but women start catching up as they pass middle age. About 55 percent of men over 60 snore, compared to 45 percent of women over 60.

What causes snoring? Bad vibrations. When we sleep, our respiration rate lowers, but we continue breathing. With luck, the air passageways between the nose and throat and our lungs remain clear while we are asleep. If so, we hear at most a few "zzzzzz's." But if there is an obstruction of any kind, the structures of the mouth (most commonly, the soft palate, tongue, tonsils, and uvula) strike against each other and vibrate, and we hear the starts and snorts of snoring.

The tongue and throat structures must be flexible so that we can create the different sounds necessary for our language, and so that we can destroy a sirloin steak and sip caramel lattes with the same apparatus. But if the air passageways from the throat down to the lungs were a stiff tube, it would serve us better for the purpose of breathing. When we sleep, our muscles relax. The muscles that help keep the air passageways in our throat open during the day tend to constrict even for nonsnorers. The soft sides of the throat pull inward, and they vibrate like a flag rippling on a windy day.

Most of the causes of snoring are not sex-specific. Anything that promotes diminution of the muscle tone in the throat can lead to snoring, which is one of the reasons why most of the sleep specialists we consulted recommend eliminating smoking, drinking, and unnecessary pharmaceuticals (especially antidepressants, antihistamines, and ironically, sleeping pills). Overuse of drugs and alcohol results in loss of muscle tone that promotes vibrations in the throat. Smoking can cause a thickening of throat tissues, shrinking the area through which air can pass freely. While there is evidence that men overindulge in all of these "vices" more than women, smoking seems to affect the level of snoring equally in both sexes.

The most common risk factor in snoring is obesity. Obese people tend to have fatty deposits located below the mucous membranes in the throat that block proper airflow. Dr. Mansoor Madani, director of the Center for Corrective Jaw Surgery, in Bala Cynwyd, Pennsylvania (http://www.snorenet.com), told *Imponderables* that many of his patients are overweight and developed snoring problems only after becoming so. Many patients report snoring disappearing after significant weight losses. Although more men than women are obese, as with the "vices" listed above, it is not enough to explain the huge disparity in snoring rates.

Snoring is one condition in which bigger is definitely not better. If snorers experience periods of total breath obstruction during sleep (usually in between snorts), the condition is known as "obstructive sleep apnea" (or OSA). The single greatest predictor of sleep apnea is neck size. If your neck size is seventeen inches or greater, you have a thirty percent chance of suffering from sleep apnea.

Obviously, many more men than women have large necks, and there is some evidence that the structures inside the throats of large people are disproportionately bulky. According to Dr. Sunny Dong, a sleep specialist in British Columbia:

> Women tend to have a shorter pharynx, which is less likely to vibrate during sleep. The soft palate is often thinner and the uvula is smaller, which makes the vibrations less loud and the frequency higher (which tends to make it less annoying).

As we age, the muscle tone in our throat tends to decrease, which results in more vibrations and snoring when sleeping. The tongue enlarges with age, which intensifies the problem. But Dr. Madani reports that the skin texture is firmer in women. Men's throat tissue gets "floppier" as they get older.

The big controversy among sleep disorder specialists is the role of hormones in snoring. Some of the experts we consulted feel that medical science still can't explain the etiology of men's higher snoring rate. Dr. Andy Blockmanis, of the Pacific Center for Sleep Disorders in Vancouver, British Columbia, speculates that men's greater weight, on average, may account for the snoring "gap," and that women's increase in snoring as they age may be attributed to their tendency to gain weight after menopause.

Most of the experts we consulted do believe that hormones affect snoring rates. They point to the obvious clue that the gender gap in the snoring rate narrows drastically after women undergo menopause, when almost as many women as men snore.

Dr. Dong points specifically to the role of progesterone, a female steroid sex hormone:

> Before menopause, women have much higher progesterone
> levels than men and this hormone increases the tone in the
> airway muscles and increases respiratory drive, both of which
> lessen the tendency of the pharynx to collapse and vibrate.
> After menopause, the progesterone levels drop and more
> women begin to snore.

Anecdotal evidence supports the case for hormonal differences as a major factor. Dr. Christian Guilleminault, cofounder of the journal *SLEEP*, feels that women are protected from snoring by progesterone and estrogens, and points out that when women are treated with testosterone, this protection often disappears.

David Nye, a physician at the Midelfort Clinic in Eau Claire, Wisconsin, notes that androgens—male hormones—increase snoring. His partner, Donn Dexter, published a paper indicating that one female patient's problems with sleep apnea were resolved after a testosterone-secreting tumor was removed in surgery.

Most of the treatments for snoring are not sex-specific. For some, eliminating vices such as smoking or drinking can solve the problems. Others find nasal sprays or air-flow masks helpful—more than 300 patents have been filed for snore-control devices. Many surgical procedures using knifes or lasers can alleviate snoring difficulties. Our favorite is "uvulopalatopharyngoplasty," a procedure to reduce the size of the uvula, the soft palate, or both.

Why would someone go under the knife for uvulopalatopharyngoplasty for a problem as benign as snoring? Ask the significant other of the snorer. Several years ago, an Iranian man filed for divorce, accusing his wife of secretly drugging him so that he wouldn't hear her snore. The wife admitted drugging her newlywed husband's dinners; the husband caught on to her trick by skipping his usual evening meal one night. The wife of only forty days volunteered to sleep during the day and stay awake at night in order to save her marriage, but the reluctant groom rejected the proposal.

Submitted by Marlena Lynch of Fort Worth, Texas. Thanks also to Neil Young of Mill Valley, California.

Why Do Vultures "Waste Time" by Circling Their Dead Prey Instead of Swooping In and Chowing Down Immediately?

We'll give you the executive summary first. When a vulture circles before eating, there are probably one or two reasons for the behavior—it is checking to make sure that the prey is actually dead or it is looking for other possible predators, especially land predators that might serve as competitors for the food. Vultures are eaters, not fighters, and neither want to kill a potential food source nor fight four-legged or winged competition for a meal.

But the executive summary leaves out a lot of fascinating information about vultures and the most shocking surprise ending since *Psycho*. There are three different species of vultures found in North America. The turkey vulture (mistakenly called buzzards by many) is by far the most common, found in all fifty of the United States, Canada, and Mexico. The black vulture generally is found only in the southeastern United States, and the California condor (despite the fancy name, it's really a vulture), nearly extinct just decades ago, can be found in the wild not only in California, but also in Arizona and Baja California, Mexico. Unless otherwise specified, we're talking primarily about the turkey vulture in this discussion.

One of the big differences between turkey vultures and their two relatives is that the turkey vulture has a strong sense of smell along with excellent vision. Another difference is that turkey vultures tend to fly lower than the other two species. These two distinguishing characteristics are likely not a coincidence, as one of the ways in which the vulture identifies its prey as dead is by smelling mercaptan, a gas emitted by decomposing carrion. The very smell that disgusts us is like the aroma of freshly baked chocolate-chip cookies to them. Ornithologists have tested vultures' sense of smell by hiding carrion; the vulture can find the dead animal just from the odor.

Vultures may be scavengers, but they do have strong preferences in food. They choose to eat herbivorous animals, such as cows and deer, over meat-eaters like dogs, cats, and coyotes, although if hungry enough, they'll settle for the less tasty carrion. They also have a strong preference for freshly killed meat over decomposed carrion, but because their beaks are relatively weak, they have problems penetrating the hides of animals. Black vultures and California condors have stronger beaks, and they occasionally "share" a meal of carrion that the turkey vulture could not polish off alone (it is far more likely, though, that the black vulture and condor will chase the turkey vulture off the prize, without so much as a finder's fee as recompense). More often, the turkey vulture bides its time until it can devour the carrion itself.

Ready for the surprise ending? Most of the experts we consulted did not agree with the premise of the question. One turkey vulture expert, who prefers to remain anonymous, rather angrily pointed us to the FAQ (frequently asked questions) section of the Turkey Vulture Society's Web site, which clearly states: "Contrary to popular belief, vultures do not circle over dead or dying animals."

How can we account for the "myth," then? To conserve energy, vultures flap their wings as little as possible. They can soar for great distances and long periods of time by taking advantage of thermals—updrafts of rising, warm air. Because of the birds' less than sterling reputation, human bystanders might assume that the vultures are circling prey, but turkey vultures often circle together in thermals just to gain altitude—if they catch the drafts properly, they can soar for hours without flapping their wings.

Black vultures are much more social than turkey vultures. When they encounter large carrion, they often fly above the prey to attract fellow vultures to help polish off the carcass—the behavior isn't necessarily altruistic, as there is safety in numbers. The black vultures are well placed to see turkey vultures below them, and can pick off a discovery from a turkey vulture. One turkey vulture has a good chance to fend off a single black vulture, but not a flock of them.

Whether we call it soaring, circling, or hovering, the vulture is careful to make sure its prey is dead before hitting the ground, and careful to look for competition from other birds and animals. But all experts agree on one thing—once the vulture thinks that all signals are go, it doesn't dilly-dally: it swoops in quickly and chows down!

Although vultures may have many competitors for food, they have few natural predators, which is lucky for them, for they have few weapons in battle. Their best shot at deterring enemies is a direct result of their less than appetizing diet, as Sy Montgomery, in his article, "Heavenly Scavengers" in *Animals* magazine, explains:

wild vultures have habits some people find, well, unsettling. Their method of self-defense, for example, is to vomit their food, which they can send sailing 10 feet. Remember that turkey vultures eat nothing but carrion, and the odor does not improve from the sojourn in the vulture's stomach.

The other end of vultures can be hazardous, too. Their white legs only look white; it's really "whitewash" from the material they defecate on their legs. [The waste matter contains acids that help kill bacteria from the carrion that results from the vulture's legs coming in contact with the decomposing carrion.]

Though these behaviors might distress people, they serve turkey vultures well. Vulture vomit is an effective predator repellent, as researchers who have worked with the species can attest.

Submitted by Susan Friend-Trommatter of Surry, Virginia.

What's the Purpose of the "SysRq" Key on Most Computer Keyboards?

The "SysRq" key, short for "System Request," is one of those features added by computer geeks to let us average users know that we are but mere interlopers in their world. The good news is that pressing the SysRq key (sometimes labeled "Sys Req") by mistake is unlikely to do any damage to your computer session. The bad news is that pressing the SysRq key is unlikely to do anything at all.

On most PC keyboards, the SysRq key shares a key with the Print Screen command, which has the virtue of actually doing something (creating screen shots of what you see on your monitor), but usually only if preceded by pressing Ctrl+Alt. In other words, if you are word processing and mistakenly hit the Print Screen/SysRq key, nothing seems to happen (in some systems, hitting Print Screen will make a copy of the screen on your clipboard).

At one time, IBM had great hopes for the now obscure SysRq key. It appeared as a function key on the keyboards used to control IBM's popular 3270 terminals (designed to interact with IBM's mainframe computers). SysRq allowed users to directly communicate with the underlying operating system to, say, switch sessions on the host server. Circumstantial evidence indicates that IBM had the same role in mind for the key for personal computers, as SysRq did not appear on the 83-key PC/XT keyboard but became the 84th key on the AT keyboard (designed to work with Intel's then-snazzy 286 processor) in 1984. But most programs never took any advantage of the opportunity to develop this "attention" key.

SysRq is making a bit of a comeback, though, as the free and open-source operating system, Linux, offers users a "magic SysRq key." Once this feature is enabled, by pressing Alt + SysRq, users can communicate directly with the underlying operating system. Once invoked, the magic SysRq key can do things like reboot the system when it is unresponsive to the usual methods or dump current memory to your console, helpful when debugging.

Although keyboards now sport twenty more keys than they did when the AT keyboard was king, most popular programs seem to utilize fewer function keys than ever (when's the last time you used the "Scroll Lock" key?). But excising a key from the computer keyboard, even if it usually lay untouched, is harder to accomplish than scrubbing a pork barrel project from the home district of a member of the House Appropriations Committee.

Submitted by Darrell Wong of Pearl City, Hawaii.

Why Do Taxicab Drivers Often Put Their Car in Park While Waiting at a Stoplight?

I t's probably not a coincidence that this question came from New York City; we've seen this curious behavior most often in big cities. Sometimes when we are confronted with a new Imponderable, we play a little game of Malarky, and try to come up with bluff answers. In this case, we could think of at least three cool alternatives:

A. If you're stopped safely at a stoplight and put the car in Park, the car can't move accidentally. It's done for safety reasons.

B. Putting the car in Park puts less wear and tear on the brakes.

C. Putting the car in Park allows drivers to rest their legs and feet.

When we posted this Imponderable on a couple of discussion forums for cabbies, the drivers let us know quickly that the correct answer is C. They thought the answer was obvious and let us know it:

> Simple. After ten to twelve hours of driving, one's foot gets sore from holding down a pedal. Is there something wrong with that?

No sir.

While agreeing with the main point, cabdrivers pointed out another benefit of putting the car in Park. By making sure the car is stopped, it's safer and more convenient for a driver to goof around (e.g., reading the newspaper, checking out the legs on a fetching passerby, or changing the radio station) or to conduct business (particularly, paperwork necessary for their job).

As for the other Malarky bluffs, "A" makes sense, especially if the driver is distracted or is stopped on an incline. But "B" is pure hooey; constantly shifting the gears is more likely to put wear on the transmission than applying foot pressure, when stopped, is going to burn out the brakes.

And the "foot relaxation theory" explains why cabbies throw the car into Park more in big cities, where there is more stop-and-go driving—with the emphasis on *stop*.

Submitted by Alona Amsel of New York, New York.

Where Do Telephone Companies and Utilities Obtain Telephone Poles? Are They Special Logs?

They may not be much to look at, but utility poles are among the superstars of treedom. Although all logs might look alike to the layman, phone and utility companies pay several times the going price for the logs that become telephone poles, because the specifications are so strict. Although the individual utility rather than the government sets its own specifications, utility poles are expected to meet the standards set by ANSI (American National Standards Institute) and the REA (Rural Electrification Administration).

Logs are bought from outside vendors, sometimes via competitive bidding. The most common types of trees purchased are pines (especially Southern yellow pines), cedar, and fir. Specifications set by the utilities include requirements for height (a typical tree is thirty to forty feet high), width, treatment (most logs are treated to prolong their life), maximum taper (the ideal pole is as straight as possible), and the number and tightness of knots (the fewer the knots, the better).

Traditionally, logs have been treated with creosote, a liquid derivative of coal tar that has been the most popular wood preservative. The loggers are motivated to take this extra step because many utilities insist on warranties for a certain "life" of the pole (typically, twenty to thirty years). With particularly durable woods, such as cedar, sometimes only the portion of the log set into the ground is treated. As the rule of thumb is that ten percent of the log's height plus two feet need be inserted, about the first five feet of the "butt" of the log is underground. Many logging concerns now use the synthetic pentachlorophenol (PCP) as a wood preservative. Isn't PCP a carcinogen? Yep, but then so is creosote.

Submitted by Robert Holiday of Royersford, Pennsylvania.

Why Do Telephone Poles Extend Far Above the Highest Wire or Crosspiece?

Telephone poles don't *always* extend far above the highest wire," Bill Sherrard, a spokesperson for Long Island Lighting assured us, although conceding that they often do. Corporate librarian Julie Swift of Michigan Bell, a division of Ameritech, backs him up:

> In the case of telephone poles, poles with only phone lines
> on them extend only about a foot above the cable to provide
> room for future expansion.

So when you're seeing a pole much higher than the telephone wires, chances are the telephone company is sharing a pole with the electric utility and, possibly, a cable operator. According to Swift, for these shared poles, "It is the electric company that determines for their own reasons how much pole to leave above the cables." Usually, the highest-voltage wires are put on top, so the order from the bottom up, is telephone, cable, and electrical wires. Of course, whoever built the pole receives rent from the other companies sharing the pole. As cable providers have inched into the telephony domain of what were once monopolies, some telephone companies have fought back by trying to drastically raise their rates.

Although the main purpose of the "extra" height of the pole is to provide for expansion, there are at least two other benefits provided. The tops of most poles are grounded so that if they are hit by lightning, the charge will hit the ground wire rather than disabling a working circuit lower down. And the excess pole gives birds an alternative to crosspieces as a place to perch. The danger isn't from electrocution by perching on copper wires. The wires are well insulated, and even if they weren't, a body is a lousy conductor compared to a wire—no current would flow through the bird. But when a big bird spreads its wings and accidentally touches two energized parts or one energized wire and a grounded metal part, the result is one sizzled bird.

Submitted by Allen Jamieson of Sacramento, California.

Why Does a Can of Diet Coke Float in Water While a Can of Coca-Cola Sinks?

At first, we didn't want to answer this Imponderable. For some reason, this is a favorite question among listeners on call-in radio shows, but we wondered whether many people really ponder over this conundrum. When a caller on a radio show recently sounded downright desperate for an answer, we decided to do something about her plight.

We filled up the official Imponderables Central bathtub with water and proceeded to gently place a can of Diet Coke and a can of Classic Coke, both twelve-ouncers, into the tub. Indeed, the Coke dropped like the proverbial stone, while the Diet Coke floated lugubriously on the surface.

As is our wont when we are faced with a daunting physics challenge, we sidled up to our consulting physicist, John DiBartolo, of

Polytechnic University (who will henceforth be identified as J.D.) and demanded an explanation:

> **IMPS:** *What's the deal?*
> **J.D.:** *The upward force on a submerged can is the buoyant force, and it depends on the can's volume (its size). The downward force is gravity, or in other words, the can's mass (its weight). (Mass is a property closely related to weight. Although technically not the same, for the purpose of this discussion we can use these terms interchangeably.)*
>
> *If Diet Coke floats while Coke sinks, it means at least one of the following:*
>
> *1) A can of Diet Coke is larger than a can of Coke*
> *2) A can of Diet Coke weighs less than a can of Coke*
>
> *I'm not sure which of these is true.*

So, we contacted our pals at Coca-Cola and were told that the cans used for both drinks are identical. We were not the first people to pose this question to the soft drink behemoth, and we were provided with this answer from the industry and consumer affairs department:

> Sugar-sweetened products are sweetened with high fructose corn syrup (HFCS) and/or sucrose, both of which are types of sugar. Most of our diet products are sweetened with aspartame and/or a combination of aspartame and acesulfame potassium, both of which are low-calorie sweeteners.
>
> Because aspartame is sweeter than sugar, it takes less aspartame than sugar to make a product taste [just as sweet].

Therefore, the density of Coca-Cola Classic is approximately 1.25 grams/milliliter, while the density of Diet Coke is 1.00 grams/milliliter. Based on these specifications, it would be correct to assume that the diet products would float in water, while sugar-sweetened products would sink.

Coca-Cola's statement soft-pedals the difference in the amount of sweeteners in each. There are approximately 39 grams of sugar in a twelve-ounce can of Coke, and only about 200 milligrams (one-fifth of a gram) of Nutrasweet (the trade name for aspartame) in Diet Coke. That's right—there is 195 times the amount of sugar (by weight) in Coca-Cola Classic as Nutrasweet in Diet Coke. So now we ran back to Mr. Physicist:

IMPS: *The labels say there are 12 fluid ounces and 355 milliliters in each can. Do they mean the same thing?*
J.D.: *Yup. You know that 355 milliliters refers to fluid volume because a "milliliter" is a unit of fluid volume. A "fluid ounce" is simply another unit of fluid volume. (One fluid ounce is 29.6 milliliters).*

The volume of something is a measure of the amount of space it takes up. 355 milliliters of water and 355 milliliters of mercury both take up the same amount of space.
IMPS: *If there is more sugar in Coke than Nutrasweet in Diet Coke, does that mean there is less unsweetened liquid in Coke than Diet Coke? Does that mean Coke weighs more than Diet Coke?*
J.D.: *Since the complete recipe for each beverage has to fit in 12 fluid ounces of volume, then more sweetener means less space for presweetened liquid. Since sugar and Nutrasweet are each denser than the presweetened liquid, any space taken from the presweetened liquid and given to the sweetener will increase the overall*

weight of the sweetened beverage. (Actually, a certain mass of a sweetener takes up less space when dissolved in a liquid than it does when in its solid form. That means that it does a particularly good job at increasing the mass of the sweetened beverage while taking up very little space.)

This means that the same volume of Coke (which has much more sweetener in it) weighs more than the same volume of Diet Coke. In other words, Coke is more dense than Diet Coke.

IMPS: *OK, so Coke is denser than Diet-Coke. Why does that make Coke sink?*

J.D.: *The heavier an object is, the greater the downward pull on the object. The bigger the object is, the greater the upward pull on the object. When an object is denser than water, the downward pull is greater than the upward pull, and the object sinks. When an object is less dense than water, the upward pull is greater than the downward pull, and the object floats.*

IMPS: *So if you have two identical cans, does it matter what's inside as far as "sinkability" goes? Will a 12-ounce can of Coke sink and a can of marbles that weigh less float? Does it matter what kind of liquid is in the can? Whether it's carbonated or not? Or is weight (its mass) all that matters?*

J.D.: *The only thing that affects whether the can floats or sinks is the weight of its contents. This is because the only two determining factors for buoyancy are can size (unaffected by its contents) and can weight (affected only by the weight of its contents).*

Therefore a can will behave exactly the same way if it contains Coke, marbles, or tiny marshmallows, provided the weight of each of these is the same.

IMPS: *Prove it, physics boy!*

J.D.: *OK, Imponderable dude. I weighed six different cans of Coke as well as six different cans of Diet Coke. I then emptied the* *cans and repeated the measurements. For each six-pack, I sub-tracted the average empty can weight from the average full can weight, which gives the average weight of the fluid inside the can. Here are the results (together with the error due to instrument precision and statistical variation):*

Mass of fluid in can of Coke: 370.5 grams (± 0.8 grams).
Mass of fluid in can of Diet Coke: 353.8 grams (± 1.3 grams).

Based on these measurements, it appears that the contents in a can of Coke weigh about 17 grams more than the contents in a can of Diet Coke.

IMPS: *And yet they don't charge more for the Coke!*

J.D.: *Shhh, we're conducting an important experiment here. Now that I showed that Coke weighs more than Diet Coke, the next question is: Is this true because there's more fluid in the Coke can than in the Diet Coke can? To answer this, I measured the volume of the liquid in each can. As it turns out, when Coca-Cola says there are twelve fluid ounces (355 milliliters) of soda in each can, they're not kidding.*

My measurements showed that for both Coke and Diet Coke, the average volume of the fluid inside was indeed 355 milliliters (±0.5 milliliters). Because the volumes are equal for both drinks, this tells us that the discrepancy in weight is due to the fact that Coke is more dense than Diet Coke (where density is found by di-viding mass by volume). Speaking of densities, here are the results:

Density of Coke: 1.044 grams/milliliter (±.004 grams/milliliter).
Density of Diet Coke: 0.997 grams/milliliter (±.005 grams/milliliter).

As we guessed earlier from the large difference between masses of sweeteners added to each drink, there is a difference in drink densities. (The density of Coke quoted by Coca-Cola seems to be way off, by the way.) Well, whaddya know? This physics thing actually works sometimes.

Is the Nobel Committee watching? The 2005 winners in physics won for ". . . contribution to the quantum theory of optical coherence" and "the development of laser-based precision spectroscopy, including the optical frequency comb technique." We think DiBartolo's Coke–Diet Coke research is more important.

Truth be told, we figured out the density part of the equation, but it didn't occur to us that in a sense the density issue is a red herring: The greater amount of solids (i.e., sugar) makes the Coke denser than Diet Coke; the greater density of the Coke makes the full cans of Coke heavier than the Diet Coke; but ultimately it is only its higher weight that sinks the Classic Coke.

Submitted by Jay Ballinger, of parts unknown. Thanks also to Craig Blanchard of Seattle, Washington; and Shana Grey, via the Internet.

OUR TOP SKATER HAS DONE A FLAWLESS QUAD! AND, TABLE #2, I'LL BE RIGHT OVER WITH YOUR CHATEAUBRIAND, OK?

Why Do Male Figure-Skating Announcers on Television Wear Tuxedos?

D o swimming commentators wear Speedos? Do wrestling announcers wear singlets?

Yet male figure-skating announcers look more like they're going to the prom than to a sporting event. For that matter, the female commentators often look like their prom dates. What's the deal?

According to Carole Shulman, executive director of the Professional Skaters Guild of America, a clue lies in the lineage of the sport:

> Figure skating is an elegant and sophisticated sport. It has been enjoyed by royalty dating back as early as 1660 to the court of Charles II, Duke of Monmouth. Beautiful clothing

including long skirts adorned with fur worn by the ladies and waist coats and top hats worn by the gentlemen, can be seen in prints depicting early scenes on the skating pond and mentioned in Scandinavian literature dating back to the first century.

At the turn of the twentieth century and until the 1960s, clothing styles changed but the predominant competition costume for male skaters was the tuxedo. This was the competition era from which Dick Button emerged, first as an Olympic champion and later as a television color commentator. It was natural for Mr. Button to continue the tuxedo from the ice to behind the microphone.

While we weren't able to contact Mr. Button, in 1999 we spoke to the late Ronnie Robertson, who won the silver medal in the 1956 Olympics. He agreed with Shulman that Button, Sonja Henie, and the other superstars of their era were the successors not only to the elegant European tradition, but the tony early days of U.S. skating, dominated by private skating clubs in Boston, Cleveland, Detroit, and New York. Robertson observed that these clubs

were very exclusive and powerful in the United States Figure Skating Association and it was doctors and lawyers whose families belonged and whose children competed. It was considered a rich man's sport.

Figure skating has always straddled the line between entertainment and sport. Recently, in a *New York Times* interview with Guy Trebay, 1984 gold medalist and television commentator Scott Hamilton bemoaned some of his fashion atrocities:

"You look at all the beading and the sparkles, and the cringe meter goes to the red," said Mr. Hamilton, who attributes the current stylistic nadir to the influence of television, which made skating "more about show business and theatrics and less about athleticism" and, oddly, to an expanding fan base for the sport.

The problem is that the sparkly, spangly jump suits that were briefly trendy in the early '80s look as dated as the disco apparel from the 1970s. Dick Button had the right idea. Scott Hamilton is looking much, much better in his conservative tuxedo.

Submitted by Marian Stoy of Hi-Nella, New Jersey.

Why Is It More Tiring To Stand Than To Walk? Why Is It Hard To Stand Still on Two Legs?

We're quite fond of moving as little as possible. As far as we're concerned: "La-Z-Boy Recliner, *sí*; exercise, *no*!" But we've noticed the same phenomenon as our readers. When forced to stand still for prolonged periods, we feel as antsy as three-year-olds who have drunk too much Pepsi-Cola and had no access to a bathroom.

While there might be some psychological elements at play, the physiological forces are probably dominant. When we stand, gravity pulls the fluid in our body down. The fluid can then pool in the legs and feet, putting pressure on the muscles, and eventually, even pain.

When we move, the contraction of the muscles pushes the fluids back up, and the veins contribute by sending blood back to the heart.

247

Walking increases the circulation not just to your heart, but all over your body, including to not-insignificant organs, such as your brain.

Even sitting for a long time can cause circulation problems. As we discussed in *Why Do Clocks Run Clockwise?*, swelling feet (edema) can be traced to passengers' lack of movement and sedentary state on airplanes. In extreme circumstances, passengers can suffer from DVT (deep vein thrombosis), when a blood clot develops. That's why most physicians recommend walking around a bit on long flights, or at least stretching your calves periodically—which might be safer than dodging the service carts in the aisle.

Airplane seats force your legs to be perpendicular to the floor, so it's easy for fluid to build up, still another reason why we're antsy on flights. To the extent that antsyness while standing or sitting straight causes us to fidget or move, our discomfort is serving an evolutionary advantage. But the La-Z-Boy devotee has the right idea. By propping up your feet while you sit, you're not only catching more zzz's, but preventing edema.

Submitted by Gerald S. Stoller of Spring Valley, New York. Thanks also to Dot Finch of Soddy-Daisy, Tennessee; and Claudette Hegel Comfort and Bob Parker of Minneapolis, Minnesota.

If Thanksgiving Is a Harvest Celebration, Why Is It Held in Late November?

The Pilgrims contended with horrendous conditions in freezing New England in the early seventeenth century—blizzards, famine, and epidemics that decimated their population were no picnic. So perhaps we needn't needle them for picking a time to give thanks for Nature's harvest when about the only harvestable crop was snow cones.

Actually, the accounts of the first Thanksgiving are remarkably sketchy—all we know is based on two written accounts by colonists. The most detailed account was written by Edward Winslow, three-time governor of the Plymouth colony, who said that the first celebration took three days. The fifty-two surviving colonists invited "some 90" Wampanoag Indians to the feast, who proceeded to thank the Pilgrims by killing five deer, "which they brought to the

Plantation and bestowed on our Governor, and upon the Captain and others." Winslow makes clear that the celebration was in honor of the harvest, and that they dined on "fowls" (unspecified in nature, but almost certainly not turkeys). The second account, which corroborates most of the information in Winslow's letter, was written by William Bradford (another Plymouth governor), in a book written twenty years later.

Significantly, neither version referred to the harvest celebration as "Thanksgiving" nor indicated the exact date of the festivities. Most likely, the colonists were heartened by their first bountiful crop, and were inspired by traditional English harvest celebrations. If they thought they were starting an important new holiday, they probably would have repeated the revelry the following year—but they did not. Historians at the Plimouth Plantation believe that the three feast days occurred sometime between September 21 and November 11, 1621.

Chances are, the Pilgrims would have been dismayed by the appropriation of the word "Thanksgiving" to describe their three-day party. To the Pilgrims, thanksgivings were days of solemn prayer and contemplation in church, not festivities featuring eating, singing, dancing, games, and merriment.

Over the next couple centuries, a series of proclamations declared official holidays of Thanksgiving. The first attempt to make a holiday of gratitude was in 1676, when the Massachusetts council proclaimed the balmy date of June 29 as a day of Thanksgiving, although obviously not a harvest festival. The first time all thirteen of the original colonies celebrated on the same date was in 1777, to commemorate the victory over the British forces in Saratoga. George Washington proclaimed a national day of thanksgiving; but soon Thomas Jefferson yanked it away. Up until the Civil War, Thanksgiving celebrated different things at different dates in different places.

DAVID FELDMAN

Before the Civil War, most communities celebrated local harvest festivals. The dates of the festival tended not to be fixed, as harvest dates in the same locale varied from year to year (farmers didn't feel like celebrating before the crops had been picked). Most local Thanksgivings were held between mid-September and mid-October, just after crops had been harvested.

You'd think that the middle of the Civil War wouldn't be an opportune time to launch a new holiday, but then it would be hard to conceive of someone as obsessed with the subject as Sarah Josepha Hale, a novelist turned media star, who became the editor of the influential women's magazine, *Godey's Lady's Book*. Hale used her soapbox to create a mythology about the first Thanksgiving that lives on today—that the Pilgrims supped on plump, stuffed turkeys, and polished off the meal with pumpkin pie. But Hale didn't stop with Martha Stewartish features about Pilgrim cuisine. She used her editorial page to lobby for a national Thanksgiving holiday, and privately lobbied politicians and other prominent people.

The American Civil War lasted from 1861 to 1865. Why did Abraham Lincoln declare Thanksgiving a national holiday in the *middle* of the war, 1863, when he must have had much more pressing matters on his mind? Although Hale was relentless in her pressure, she had been railing at presidents for decades without success.

May we bring up the dreaded word, *politics*? After many victories, the Union encountered a series of reversals in 1863. Robert E. Lee's troops were wounded but not routed at Gettysburg, and then Major General William Rosecrans's troops were crushed by Confederate forces at Chickamauga Creek in Tennessee. More than 35,000 Union soldiers were lost at Chickamauga, and Lincoln referred to Rosecrans as "confused and stunned like a duck hit on the head."

While spirits were low and casualties were high, Lincoln had something else to worry about—reelection. In his Thanksgiving

Proclamation, Lincoln alluded to the war as being "of unequaled magnitude and severity," but parts of it read like a political campaign speech:

> peace has been preserved with all nations, order has been maintained, the laws have been respected and obeyed, and harmony has prevailed everywhere except in the theatre of military conflict; while that theatre as been greatly contracted by the advancing armies and navies of the Union. Needful diversions of wealth and of strength from the fields of peaceful industry to the national defence [sic], have not arrested the plough, the shuttle, or the ship; the axe had enlarged the borders of our settlements, and the mines, as well of iron and coal as of the precious metals, have yielded even more abundantly than heretofore.

The second and last paragraph of the Proclamation is uncompromisingly religious. Lincoln was consciously detaching "his" Thanksgiving from its agricultural roots both by moving the holiday past harvest times and by giving the national holiday a religious justification that also attempted to soothe the wounds of war for both North and South:

> And I recommend to them that while offering up the ascriptions justly due to Him for such singular deliverances and blessings, they do also, with humble penitence for our national perverseness and disobedience, commend to his tender care all those who have become widows, orphans, mourners or sufferers in the lamentable civil strife in which we are unavoidably engaged, and fervently implore the interposition of the Almighty Hand to heal the wounds of the nation and to

> restore it as soon as may be consistent with the Divine pur-
> poses to the full enjoyment of peace, harmony, tranquility
> and Union.

Lincoln was likely sincere in his comments, but he was also a politician, up for what would probably be a bitter election in one year. He
wrapped his proclamation in the trappings of religion in a way that
would have horrified the Pilgrims.

Although a national Thanksgiving continued to be celebrated
on the last Thursday of each November after Lincoln's assassination,
it was not officially a national holiday—technically, each Thanksgiving was proclaimed by the sitting president annually. Thanksgiving
didn't become an official national holiday with a set date until another cataclysmic event—the Great Depression. FDR decided that
rather than being celebrated on the last Thursday of November, it
should be on the fourth Thursday of the month. Most years, the
two dates would be the same, but in 1939, there were five Thursdays
in the month, just as there were earlier in his term, in 1933. In that
earlier year, when the economy was in even worse shape, the business
community lobbied Roosevelt to move up the date of Thanksgiving,
because most Christmas shoppers waited until after Thanksgiving to
start spending on gifts. During the Depression, businesses needed
every break they could get. Roosevelt resisted their pressure in 1933,
but acquiesced six years later.

FDR's decision was met with all kinds of abuse. Traditionalists
didn't want to change the long-held custom. Schools and some businesses didn't have enough time to change their vacation schedule.
Calendar makers weren't given enough lead time to make the
change not just for 1939, but for 1940. All this was bad enough, but
FDR was even messing up football schedules. The mayor of Atlantic
City, New Jersey called the rescheduled holiday "Franksgiving."

Many states refused to comply with FDR's edict, which caused further problems, as families from different states couldn't meet on a Thursday to chow down on the turkeys that the Pilgrims never ate. Twenty-three of the forty-eight states adopted November 23, 1939 as the date for Thanksgiving, while twenty-three refused to alter from November 30; two enlightened states, Colorado and Texas, honored both dates. In 1940, a few more states went along with the fourth-Thursday scheme, but finally, in 1941, Congress took the power out of the presidents' pens and officially declared Thanksgiving to be on the fourth Thursday of November.

And so it stands, until, perhaps, another crisis comes along. The weird timing of Thanksgiving has much to do with economics, politics, religion, and tradition, and little to do with agriculture or the Pilgrims.

In a *Chicago Tribune* column, aptly dated November 25, 2005, Eric Zorn chronicles some of the oddities we have discussed, and tries to spearhead a return of the holiday to its Pilgrim roots, in October, but his motives are not just to honor history, and he's willing to take on his state's most illustrious politician in the process:

> Thanksgiving in October would mean no need to surf the Web on Saturday evening wondering if you'll make it back home the next day or if you'll spend Sunday night sleeping on an airport cot or in the median of the interstate where your mini-van finally came to rest.
>
> Lincoln didn't know from airports or interstates, but what's our excuse for perpetuating his mistake?

Submitted by a caller on the Mike Rosen Show, *KOA-AM in Denver, Colorado.*

DAVID FELDMAN

Why Do Self-Service Gas Stations Usually Disable the Automatic Handle on Gas Pumps?

When fate takes us to the turnpikes of New Jersey, a state where service stations, by law, must provide full service, we always admire the gas jockeys at the rest stops. They set the automatic clip on the gas pump handle while fueling and run around from car to car, free as birds. But when we stop at a self-service station, we are stuck with our hands fastened to the pump handle tighter than a Jennifer Lopez Oscar dress. Why is there discrimination against us self-service habitués?

We contacted two major oil companies and they passed the buck rather quickly. "It's not up to us," they replied. As Don Turk, a spokesperson for Mobil Oil Corporation put it, "The general rule is that the use of these devices is regulated by the local or state fire codes."

Sure enough, automatic fuel clips are outlawed in many states. The National Fire Protection Association endorsed the safety of particular brands and models, and in those states where clips are legal, it is common for these models only to be permitted. But even in states where clips are allowed, relatively few service stations use them in practice. Gasoline is expensive enough—why do service stations have to torture us with manual labor?

We asked this question of Paul Fiori, executive vice president of the Service Station Dealers of America, and he wouldn't buy conspiracy theories. He wasn't sure, but speculated that fuel clips tend to break, some customers don't know how to use them properly, and the end result is usually more expense for the dealer.

Citgo service station owner Maurice Helou, of Lyndhurst, Ohio, backs up Fiori:

> I have been a gasoline dealer for 25 years. There are many
> procedures a self-service gasoline customer should practice
> that may result in a request from the cashier at the store,
> such as "pull up to the next pump," "no smoking please,"
> etc.
> People do not like to follow instructions, requests, or com-
> mands from the "lowly cashier." Since the gasoline nozzle and
> its assembly is a piece of equipment subject to failure and mis-
> use, the result of which would cause a spill, by disabling the
> clip, the consumer is forced to control the flow of gasoline
> into their tank, avoiding a costly spill and its resulting com-
> plaint from the consumer.
> Most fire departments require the customers to control the
> gas nozzle. Take away the clip and that's one less request.

Bah humbug! In our experience, the shutoff mechanism always works, and at a self-service station, it's still up to the consumer to place the nozzle of the pump into fuel tank. Based on some of the folks we've seen at gas stations, we'd trust an inanimate clip more than humans to prevent spills. When we challenged Helou with these assertions, he swatted them away with gusto:

At the tip underside of the nozzle spout there is a small hole. When the gas tank is full, gas backsplashes into the little hole, which triggers the gas nozzle to shut off and release the clip, therefore stopping the flow of gasoline. If there is a malfunction anywhere in the process, the nozzle may not shut off and gas will continue to flow. That's why it is necessary for the customer to remain with the pump when filling the tank.

It wouldn't be necessary to remove the clip and force the customer to stay with the pump if they would simply comply. Customers are time-starved and multitask when buying gas. They set the nozzle to pump their gas and then they use the restroom, or buy products in the store while leaving the pump unattended.

When a spill occurs, then it is a hassle to collect the sale. Even when customers are manning the pump, if they attempt to "top off " after the nozzle automatically shuts off, they can overfill and cause a spill.

We spoke to Pat Moricca, president of the Gasoline Retailers Association of Florida, who confirmed that the spill issue was the paramount reason for disabling automatic fuel clips. If enough gasoline is spilled, the HazMat crews from the local environmental agency must be called, leading to lost sales. If a customer trots in with a lit cigarette, the consequences can be worse.

The reason why you see automatic clips used at full-service pumps is because the service station worker can wash the windows or check the air pressure of tires while fueling, but if there is a malfunction, he's in the vicinity to troubleshoot. Ultimately, the disabling of fuel clips is there to force the customer to stay with the car.

And remember Paul Fiori complaining about fuel clips breaking? What causes this problem? According to Louis F. Ferrara Jr., a Sunoco station owner in Philadelphia, Pennsylvania, the fault lay not with the clips, but with the customers who engage them:

> Though it sounds crazy, customers forget that the pump handle is in the gas tank. They pull the car away, causing damage to the pump. We'd rather they hold the handle until they are finished.

Submitted by Jeffrey Joyner of Raleigh, North Carolina.

Can Women Use "Just For Men" Hair Coloring?

This Imponderable was posed by Marty Flowers, a woman with a name usually associated with men. Perhaps her experiences as a female Marty made her wonder:

What happens if a woman uses it? Is it really just for very short hair? If a woman has really short hair, could she use this product successfully, or would she hate the results? If a guy has long hair, should he not use this product? So shouldn't it be called "Just for Short Hair"?

You've got the right idea, Marty. Just For Men comes from Combe Incorporated, the same folks that bring you Grecian Formula. If you care to blaze new trails, Marty, feel free to try Just For Men. It won't do any damage, except to eradicate gray hairs, just as it's safe for men to use Secret deodorant (despite the harrowing knowledge that Secret was "made for a woman"). Would it hurt men to smoke Virginia Slims? Whoops, bad example. But at least cigarettes are equal opportunity offenders—they will harm both men and women!

Grecian Formula does not contain dye and compensates for the lost melanin that results in white or gray hair. It is applied over many weeks, and the user's hair darkens gradually over time. The melanin in black hair is no different in hue from a blond's melanin—there's just more of it in the dark hair (so chances are, it will take longer for Grecian Formula to restore a black-haired user's natural hair color than it would a redhead's).

Just For Men is a more conventional hair colorant, which comes in eleven different shades. It is rinsed out after five minutes and, with luck, achieves the desired result after one application.

We don't want to give the impression that the differences between Just For Men and women's coloring products are nonexistent, but to emphasize that Just For Men's advertising and marketing is clearly designed to assuage men's defensiveness and skittishness about using coloring products. Stress is put on the elimination of gray hair rather than the change of color. Like Grecian Formula, Just For Men "targets the gray" rather than trying to convince men to frost their hair or change from brunette to blond. Many women's coloring agents are marketed as fun fashion statements, even ones meant for use on a special occasion. But Just For Men is aimed squarely and dourly at getting out the gray and retaining the original hair color, as Ralph Marburger, marketing director for men's haircolor at Combe Inc. explains:

> Most men who think about using a hair color just want to
> get rid of their gray hair and not change the color of the hair
> that has not gone white yet, because they want the result to

look like their own hair color. Just For Men is a "Deposit-Only" Hair color, which means that it does not bleach out hair's natural pigments and then deposits color molecules—it just deposits color. Since it does not bleach the hair, the only visible change to the user's hair color is the change from white hair to colored hair. The color of the previously pigmented hair stays virtually unchanged (unless, of course, someone with light hair uses a very dark shade).

How else is Just For Men designed for guys? Let us count the ways:

1. Just For Men is placed in the shaving and grooming aisles of stores, rather than in the land of Clairol and L'Oréal, an area as baffling to most men as the cookie section of the supermarket is to Marilu Henner. Subliminally, this also assures men that use of the product is part of their grooming and not a flight of vanity.

2. Men are not usually as patient as women when using cosmetics. Just For Men takes about two minutes to comb in and about five minutes to set, compared to the usual 30 to 40 minutes for women's packaged hair coloring products.

3. Unlike many women's products, Just For Men contains no ammonia, which, according to Marburger, "can damage hair and smell bad."

4. Although Just For Men isn't advertised specifically for men with short hair, there is less coloring in a Just For Men bottle than most women's dyes.

5. Combe also offers a Just For Men product for eliminating the gray in sideburns, beards, and mustaches, designed especially for coarse hair.

Never one to cede dominance in any area of hair coloring, Clairol has introduced Natural Instincts for Men (nothing says "Natural" more than artificial coloring, evidently), the only direct competitor to Grecian Formula and Just For Men in the mass market. Like Just For Men, Natural Instincts contains no ammonia and boasts reduced peroxide compared to Clairol's women's products.

Combe, playing tit-for-tat, has introduced Just 5, a product marketed to women, whose benefits and technology sound awfully close to Just For Men. We spoke to a consumer resources consultant at Combe who admitted that Just 5 hasn't caught on with women as they had hoped. Perhaps women cannot believe that a product that works so quickly can be effective? No time drain, no gain?

That's why we're happy to announce that some women are enterprising enough to cross the gender divide and use Just For Men. According to the same consumer resources consultant at Combe, they sure do, especially for touching up gray hair at their temples. As far as we know, none of these women spontaneously combusted.

Submitted by Marty Flowers of Weirton, West Virginia.

Drug Labels List Active and Sometimes "Inactive" Ingredients. What Are Inactive Ingredients and Why Are They Used If They Don't Do Anything?

S itting next to the Imponderables Central keyboard is a bottle of CVS "generic" ibuprofen tablets. Under "Drug Facts" on the label, it states that there is one active ingredient: Ibuprofen 200 mg. Coming after sections called Uses, Warnings, Directions, and Other Information, is a list of Inactive Ingredients: colloidal silicon dioxide, corn starch, croscarmellose sodium, hydroxypropyl methylcellulose, iron oxides, microcrystalline cellulose, stearic acid, and titanium dioxide. As the U.S. Food and Drug Administration defines it, active ingredients are any component of a drug that is intended to

furnish pharmacological activity or other direct effect in
the diagnosis, cure, mitigation, treatment, or prevention of
disease, or to affect the structure of any function of the body
of humans or other animals.

Inactive ingredients are *any* component of a drug product that isn't
an active ingredient.

The FDA has approved almost 800 chemical agents as inactive
ingredients for drugs, and although these inactive ingredients must
be labeled for topical preparations and eye treatments, no such requirement is issued for drugs taken orally. Most of the major phar-
maceutical companies list the inactive ingredients, but even some of
these companies leave out "trade secret" components that may tip
other companies to a competitive advantage of a particular drug.

If inactive ingredients don't help in the treatment of the condition of a patient, why are they included at all? Most commonly,
they add to ease of manufacture of the drug, the stability of the finished pill or liquid (so that the tablet or capsule holds together), and
its palatability to the user. Some ingredients, such as starches, are often used just as a filler, to keep the components of the drug solid.
Dyes are added to improve the appearance of the drug, and sometimes to give it a distinctive brand identity. Inactive ingredients, also
known as "excipients," are designed not to interfere with the delivery of the active ingredients to the end user.

The ingredients included in our ibuprofen tablets turn out to
have these benign purposes. The scary sounding croscarmellose
sodium is not toxic, and helps tablets to dissolve in the stomach. The
silica and stearic acid are used as lubricants so that the tablet doesn't
stick to manufacturing equipment. Cellulose is used as a binder to
hold the tablets together, as well as filler. Titanium dioxide is used
as a thickener.

DAVID FELDMAN

But not everyone is so sanguine about the safety of inactive ingredients. The American Academy of Pediatrics has long lobbied for compulsory listing of all inactive ingredients on prescription and over-the-counter drug labels. Pediatricians report that saccharine and aspartame, used as sweeteners in chewable tablets for children, can induce headaches. Some children respond adversely to dyes found as coatings of pills; without labels, parents can't easily tell whether their children will react.

Adverse reactions to inactive ingredients are not confined to children. About 20 percent of all drugs contain lactose (milk sugar), used as a filler or a diluting agent in tablets and capsules, and to provide bulk to powdered remedies. Those with dairy allergies or extreme lactose intolerance can suffer from side effects worse than the conditions the pills are meant to treat. Many folks are allergic to corn, and might be shocked to learn that their adverse reaction to some drugs might be caused not by the active ingredients, but by starch used as filler.

So despite the adjective "inactive," one of these ingredients can "get busy" for the unlucky few. If you worry about this happening to you, consult the *Physicians' Desk Reference,* which usually has a complete listing of inactive ingredients.

Submitted by Scott Schuetze of Green Bay, Wisconsin.

Why Do Peanut Butter Cookies Have Crisscross Marks on Them?

We don't know where, when, or who came up with the brilliant idea of topping a peanut butter cookie with fork marks, but that won't stop us from speculating *why*. Let us consider the possibilities:

1. Watch-Out! Theory

Take us as an example. We love peanut butter. We love cookies. But we don't like peanut butter cookies. The telltale crisscross has become a warning for us to "stay away" and a convenient way for both sales clerks and customers at bakeries and pastry shops to differentiate between peanut butter cookies and other confections. Peanut allergies can be deadly, so the crisscross is the equivalent of a skull and crossbones to those so afflicted, although we doubt that allergies had anything to do with the origins of the crisscross markings.

2. Just Following Orders Theory

Recipes call for the crisscross markings and most cooks are nothing if not obedient! Jeff A. Zeak, pilot plant manager of the American Institute of Baking, adds:

> Some people put the marks on the tops because that was what someone else (Grandma, Grandpa, Mom, or Dad) may have taught them to do and they never thought as to why they were doing it.

We'd guess that this theory best explains why most home bakers adorn their cookies with the crisscross. But how did the practice start in the first place? We place our bet on . . .

3. The Fork Was In Our Hands Anyway Theory

Read just about any peanut butter cookie recipe and you will see nary a word about spoons. But forks have a way of appearing once, twice, or three times in the directions. Commercial bakers stir cookie batter by machine, but many recipes for home bakers call for the dough to be stirred with a fork rather than a spoon if a mixer isn't available. Zeak explains why:

> Some peanut butter cookie dough recipes can be quite stiff and sometimes almost dry in appearance. By using a fork to mix the dough (mashing the dough between the tines), a greater mixing action is achieved that is very much like the

action that is accomplished when using some type of mechanical mixer (with dough being forced, chopped, or smeared through the beaters).

If you look at recipes for chocolate-chip cookies, you'll see that after the dough has been mixed, you are asked to put a spoonful of batter on the cookie sheet. You needn't worry about the blob flattening and turning into a nicely-shaped finished product. Peanut butter cookie dough is not as cooperative—if left as a ball, the stiffer and stickier peanut butter dough tends not to flatten out—it'll look more like a doughnut hole than a cookie. So even if the dough is mixed by hand (a sticky proposition) or a spoon, every peanut butter cookie recipe we've seen calls for the baker to flatten the balls with the tines of a fork before putting them in the oven. Different chefs have varying techniques to prevent the batter from sticking to the fork when flattening. Zeak rolls the dough balls in granulated sugar. Others dip the fork in sugar or flour, while still others grease the fork with butter or PAM cooking spray.

Our guess is that the origins of the crisscross markings came when a chef decided that as long as the tines of the fork were required to flatten the dough before baking, why not make an artistic statement at the same time? Some bakers even add a little flair by scoring the cookies *after* they are baked.

Given the Feldman theory of housekeeping, we wouldn't be shocked if the anonymous but oft-imitated baker who invented the cross marks figured: The fork is already dirty and sticky with peanut butter—why not postpone washing the fork until the last possible moment?

Submitted by Brent Detter of Landisville, Pennsylvania. Thanks also to Cheryl-Anne Smith, via the Internet, and "Barbara," via the Internet.

HALF-MOON BAY POLICE DEPT.

629174

S. FISH

S. FISH-PROFILE

Do Starfish Have Faces?

Starfish are not fish, and experts tend to get testy if you call them "starfish," anyway—they are properly known as sea stars, and are classified as Echinodermata (spiny skinned), the same phylum of invertebrates as sea cucumbers and urchins. We tend to think of sea stars as unmoving lumps that lie on the ocean's floor, when they are actually voracious carnivores, and usually prowling for food.

It's hard to have a face when you don't even have a head. Unlike most animals that have a head, sea stars, like all echinoderms, are radially symmetrical with a top side and a bottom side, but no front or back. They feel comfortable moving in any direction, as well they should: they have five—or more—arms and absolutely no notion of forward or backward.

With the naked eye, it isn't easy to see the sensory organs of a sea star, but they have many of the skills of animals with heads. One thing they don't have is ears or a sense of hearing. And although they don't have eyes, they do have eyeholes on each arm that can sense light. Sea stars often lift an arm in order to uncover the eyespot, so they can perceive light or movement in the water. Most sea stars crave the dark, as they escape predators by taking refuge underneath or behind rocks where they cannot be seen.

Seas stars have a groove running along the bottom of each arm that contains hundreds of tiny "tube feet." These tube feet not only enable sea stars to move, but also are equipped with suction cups, which allow sea stars to grip surfaces with some of the tube feet and propel themselves forward with the others. Each arm contains a single tube foot that is longer than the other feet and does not have a suction cup. When a sea star moves, this special tube foot is able to sense chemicals in the water. Even if sea stars don't have noses, they do have a highly developed sense of smell, which comes in handy when they are seeking food—their "vision" doesn't help them much to find prey.

No eyes. No nose. No ears. No heads. Do we come up blank? We are happy to announce that they do have mouths, usually located right in the center of the bottom of the sea star.

We are not so happy to describe how they use these mouths to devour their prey. Bivalves, especially oysters, clams, and mussels, are their favorite food source, but sea stars also feast on coral, fish, and other animals that live near the floor of the sea. While it takes some skill and protective gear for us to pry open an oyster, sea stars have mastered their technique; they wrap their arms around the oyster and use their tube feet to pry apart the oyster shell. Once there is the slightest crack in the shell (one estimate is that it need be open only 1/100 of an inch), the sea star extends its jellylike stomach out of its mouth (yes, its mouth) and inserts the stomach inside the shell

DAVID FELDMAN

of the oyster. The digestive juices of the stomach move into the crack of the shell while the inside-out stomach of the sea star digests its prey. It can take twenty-four hours for a sea star to fully digest a feast of a single oyster, and all of this time the stomach is having an "out of body" experience. Only when the food is fully digested does the sea star's stomach return to its mouth. If your eating habits were this appalling, you wouldn't show your face either.

This Imponderable was submitted by two children, but the strange makeup of the sea star has inspired even experts in the field to ponder. While we were researching this question, Echinoderm scholar John Lawrence of the University of South Florida was kind enough to pass along a poem written by the renowned, late biologist from Stanford University, Arthur C. Giese. While Giese's verse might not achieve poet laureate status, and some of the vocabulary might be obscure ("sessile" refers to animals that live attached to another object its whole life, such as sea sponges), we found it charming. Here's an excerpt:

Do echinoderms have a face?

The echinoderms are the strangest race
That on our World the Lord did place.
One wonders, do they have a face?
and how kissing between them takes place.

They're built on a pentameral plan
Figure that out, please, if you can.
Well, their structures are built in multiples of fives
Instead of in pairs as in our lives.

Perhaps because in the evolutionary hassle
The original echinoderms were sessile.
And Gregory writing about the face
Says there is no face in a sessile race.

Their larvae, however, tell a different story
Because bilateralism is there in full glory
The sessile habit was a later phase
That permitted pentamerism to take place.

Please will you now at a sea star look
Or turn to the picture in your invertebrate book
You'll see not two but five little "eyes"
Looking at you and up to the skies.

Submitted by Jake Itzcowitz of Highland Mills, New York. Thanks also to Maya Itzcowitz of Highland Mills, New York.

DAVID FELDMAN

Why Do We Often Find a Folded-Up Piece of Tissue Paper Inside One of a Pair of Men's Dress Socks?

We have been mildly bemused when buying a pair of dress socks and finding an interloper inside the sock: a folded-up piece of tissue paper. Like the pins and cardboard that must be excised from new dress shirts, we've always looked upon the tissue removal as a "cost of doing business" when we forsake athletic socks for dress socks—but their presence never made much sense to us. When a couple of readers wrote in, wondering what the deal with the tissue paper was, we got cracking on solving the mystery.

Like good parents, we love all our Imponderables. But like rotten parents, we love some more than others. Although we weren't obsessed with this mystery when we started researching it, we became possessed. Right off the bat, we spoke to more than twenty people in the hosiery business and not one of them knew the answer, even folks responsible for the placement of the tissue paper in the socks.

For example, the fifteenth person we spoke to was the director of packaging for a major sock company. To protect her identity, we'll call her DoP. Here is the relevant excerpt from our conversation:

> **DF:** *I have two questions for you. Why do you put a piece of tissue paper in men's dress socks? And why in only one sock of the pair?*
> **DoP:** *Let me start with the second question first. We put the tissue only in one sock to save money.*

[We were now flush with excitement. She was the first person to give us anything but a verbal shrug. Would we finally get the answer we craved?]

> **DoP:** *As for the first part, I don't know why we put the tissue in.*
> **DF:** *Are you the person who is responsible for deciding whether to put a tissue in?*
> **DoP:** *Yes.*
> **DF:** *You're the person responsible for putting in the tissue, and you're worried about the expense, and you don't know why you do it?*
> **DoP:** *[laughing] That's a good point.*

Undaunted, we kept plugging away. One vice president at a maker of store-brand socks suggested we contact the Hosiery Technology Center in Hickory, North Carolina (the epicenter not only of the furniture industry, but hosiery manufacturing in the United States). Surely, we thought, at this august center of hosiery education, someone has done a thesis in the intimate relationship between socks and tissue paper. We received an e-mail from the director of the Hosiery Technology Center, Dan St. Louis, who said that he was stumped, but recommended I speak to a true sock guru:

> I would suggest [speaking to] Sam Brookbank, who retired from the hosiery industry with over 60 years in the business. If he doesn't know, no one does. He is 85 years old and has a mind as sharp as a tack . . . He is truly a hosiery national treasure. He has forgotten more than I ever will know about hosiery.

We called Sam immediately and posed our Imponderable, and his first response was the same as everyone else we contacted: an immediate chuckle. He paused for a bit and drawled:

> In my sixty-plus years in the business, I believe this is the first time this subject has ever come up. I don't know.

Sam gave us one clue, though. He mentioned that the tissues started popping up in the early 1960s, and definitely were not inserted into socks when he started in the business.

Still, we were crestfallen. After a few midday cocktails and reading a collection of Emily Dickinson's most depressing poems, we decided that if the Little Engine That Could could, so could we.

In desperation, we contacted our pal Gloria McPike Tamlyn, a marketing consultant for the fashion industry. She started calling her friends and before long we had a whole new list of hosiery honchos to hassle. And then one fateful day, we called a man who became our hero, Jeff Stevenson, the creative director for American Essentials, a company that is best known for manufacturing the socks sold under the Calvin Klein and Michael Kors labels. Like everyone else, Stevenson chuckled when he heard the Imponderable, but then he proceeded to discourse on the matter as if he had been thinking about the subject for hours just before we called. But his thoughts could be summarized tersely:

> It's all about the crinkling. The tissue is only in there for the sound effects it makes.

Days later, we would talk to our other savior, Larry Khazzam, executive vice president of Echo Lake Industries, Ltd., a company that makes Joseph Abboud and fine private label socks. Larry confirmed that the tissue paper has to be crinkly—he compared the appeal to the "Snap, Crackle, and Pop" of Rice Krispies. Indeed, the paper, which comes to the manufacturers in toilet paper–like rolls, is selected precisely for its high crinkliness quotient.

Both Stevenson and Khazzam mentioned that the more senses you can bring into play at the store, the more you can engage consumers. With the tissue paper, you don't influence the visual appeal of the socks, but you can influence the feel and sound. Both felt that the tissue paper added an element of luxury and refinement. Khazzam added that tissue paper is also used in other luxury garments, such as dress shirts and fine sweaters. Upscale dry cleaners sometimes add some, ostensibly to protect clothing, but mostly to impart a sensation of crispness and luxury to the cleaned clothing.

We confirmed these theories with Richard B. Gualtieri, director of men's fashion merchandising (and a former men's furnishings buyer) at New York's elegant Lord & Taylor department store:

> The tissue in the sock is there as part of the esthetic; to add a sense of luxury to the hosiery. It also adds a bit of thickness to the packaging because better men's hosiery uses finer fabric yarns, which makes them thinner than the less expensive ones.

That little piece of tissue doesn't come cheaply. According to Khazzam, it costs one to two cents per sock if inserted by machine, as it is in the United States, and three to four cents if done by hand, as it is in many foreign plants. So our director of packaging was right about why the tissue is in only one sock of a pair, and often only in one sock in multipacks of two or more pairs.

As you might have guessed, the choice of which sock is not random. The tissue is always inside the sock on the outside of the package, closest to the consumer, the one that consumers are most likely to fondle before deciding which socks to buy.

There is a little disagreement among our experts about whether the paper provides any protection for the sock at all. Gualtieri notes that socks are delivered to stores in twelve-pair prepacks, and that the paper in the outer sock helps keep the hosiery from getting wrinkled while in the box. Some fine Italian sock makers put tissue paper in the leg and foot area of the sock, which might provide more support.

Khazzam was kind enough to do some digging with his European hosiery sources, and they seem to agree with Sam Brookbank that the practice started in the early 1960s, when tissue paper was put into only the finest socks (gauge 22 and 26 socks, which are very thin). Khazzam writes:

> It was started simply as a way of distinguishing the characteristics of the more expensive socks from the rest. What makes a sock more expensive than the next is a function of the desired thickness of the sock, the gauge and the yarn raw material. The reason why the higher gauge socks cost more than the lower gauge is because the initial yarn raw material has to go through more extensive processing; special care must be taken to remove all naps and knots in the raw material; the combing process must be perfect, otherwise the slightest defect would be immediately visible to the consumer.
>
> So the practice of putting the tissue paper inside these socks differentiated the better socks from the others, and became a standard practice. The advent of the insertion machines has made it much easier to put the tissue paper inside the socks, and therefore the practice has become more indiscriminate.
>
> As to the origin of this practice, we can only guess that it began in either Italy or in the U.K., which have the oldest history of knitting the finer-gauge socks. I have gone back three generations with this question, but cannot find the exact answer.

Convinced that we were on the right track, we thought of another common use of tissue paper in an unusual setting. Serious gift wrappers almost always include tissue paper as a component in the final package. When you think about it, if a gift is already encased in a box, which itself is covered by paper wrapping, why is another layer of tissue around the gift necessary? Tissue paper is hardly the most protective covering—surely there is a sock-gift wrap connection. We contacted the king of gift wrap, Hallmark, and we hit pay dirt when Rachel Bolton, a media spokesperson for Hallmark Gold Crown Stores, answered the phone. Just our luck—Rachel has a background in gift wrap and has thought long and hard about the psychology of tissue paper!

Bolton notes that the earliest gift wrapping was probably wallpaper and tissue paper. Our ancestors saved tissue paper used in gift wrapping and treated it as a precious commodity, as paper was expensive and scarce. Nowadays, most folks toss tissue paper after the gift is opened, but some save it, as Hallmark (and other companies) has added more and more design elements to the mix (some have sprinkles, some have a shiny surfaces, some have printed graphics, and tissue paper comes in almost as many colors as Crayolas).

Just as with socks, Bolton believes that tissue paper in gift-wrapped packages is appealing because it adds the element of sound (unprompted by us, she used the word "crinkle" to describe the noise). But the tissue paper also adds a patina of elegance. The extra layer adds suspense to the gift-giving process (some folks even torture the recipient by sealing the tissue paper with a sticker, resulting in mandatory extra crinkling). The honoree feels that the gift is special and more valuable, which is also why you see tissue paper included in some sets of fine stationery and chocolates.

The more Bolton rhapsodized about the process of opening a tissue-paper wrapped gift, the more the image of a striptease occurred to us. If Hallmark sells you as much gift-wrapping product as it would like, the process of opening a gift is not unlike lifting a succession of veils. Bolton believes that adults, unlike some small children, enjoy the slow "tease" of postponing the pleasure of seeing the ultimate gift—they are "into the moment."

We're not so sure that sock purchasers are quite as caught up with their tissue paper interaction, but it has consumed *Imponderables* for a few months. How cool is it to research a practice that consumers don't understand and that most of its practitioners don't, either? Sometimes we love our job.

Submitted by Donald Montgomery of Atlanta, Georgia. Thanks also to Dan Klinge of Huntington Beach, California; and Carla Fortune of Sweetwater, Texas.

Updates

Our Inbox section is full of fans frothing about the mistakes they think we've made in our books, but sometimes they help us find new information that screams for an update. Usually, we find a particular new source of information that we're bursting to share with you. We've wanted to include an update section for a long time, and with *Why Do Pirates Love Parrots?*, we inaugurate the feature. Let us know if you'd like us to continue.

Why Do Fish Eat Earthworms? Do They Crave Worms or Will Fish Eat Anything That Is Thrust upon Them?

Appropriately enough, we addressed these burning questions in *When Do Fish Sleep?* We spoke to many experts, who emphasized that most fish were attracted to live bait that moved in the water. This year, we stumbled onto a 1994 article in *The Wall Street Journal* about a man who makes a living pondering the taste predilections of fish. Dr. Keith Jones, a biologist, is the director of fish research at Pure Fishing, the world's largest fishing tackle company. With fish tanks in his laboratory, Jones conducts empirical research on the roles of sight, smell, and sound on aquatic biting proclivities. The *Journal* article mentioned that Jones tested all sorts of variables: For example, would fish be more attracted to a lure that mimicked the torso of a crayfish without the head, or a head without the body? Jones told the reporter: "We set up behaviorial tests to let the fish make choices . . . We let the fish design the lure for himself."

What really piqued our interest is that the company's plastic worms were "a quickly growing segment." Much to our delight, more than ten years later, Dr. Jones is still pursuing his passion at Pure Fishing. When we posed Dr. Jones our Imponderable, we were pleased to find that this Ph.D. is concerned with the important things in life:

> Strange that you should ask me about why fish eat worms, because I have long pondered that same question. Specifically, I'm puzzled by the fact that as terrestrial, not aquatic creatures, worms—at least earthworms—are viewed by fish as food. Worms live in the soil, not water. About the only time they ever make it into the water is when they get accidentally washed down by strong rainwash. Thus, a worm's presence in the water is not the norm, but novel. Most fish go through their whole lives without ever seeing one. And yet, despite the worm's novelty, a wide variety of fish not only readily savor the flavor of worm, they attack worms (and soft plastic worm-shaped baits) with fervor.

We noted in *When Do Fish Sleep?* that fish, like most animals, tend to prefer prey common to their environment. Jones concurs:

> Predators are generally designed . . . to successfully feed on their common prey (otherwise, the predator won't be living long), so it makes sense to me that a bass would attack a minnow even when its their very first minnow encounter. To a degree the bass is neurologically designed or predisposed to

attack minnows because bass are phylogenetically fitted to
their food . . .

About the only reasonable answer I have been able to come
up with is that bass (and other piscivorous [fish-eating] fish)
somehow mistake worms for long, skinny baitfish. If that's the
case, then one could predict that perhaps bass are not entirely
satisfied with their earthworm experience.

So if Jones's theory is correct, earthworms are the victims of mis-
taken identity! He speculates that the earthworm's shape might be
just on the fringe of the shape profile that a bass would attack—and
in his tests with artificial worms, he proved that the average worm
would be more attractive to a bass if it were shorter and plumper—
more fishlike.

Jones confirms that smell definitely plays a part in attracting a
fish. Earthworms

constantly produce a mucous covering for their skin. The
mucous is water soluble and, apparently, has an attractive
smell to fish. However, worms become much more attractive
when they are pierced with a hook, causing their internal fluids to spill into the water. The same would be true of minnows, insect larvae, etc.

Pure Fishing (and its competitors) adds scents to their lures. In the
past, liquid attractants were applied to the outside of the bait, but
these tended to wash off when they were dunked in the water. Now,
Pure Fishing also offers baits with scent inside, so that the scent
exudes from within.

Different fish have different skills in detecting prey, and different preferences, too. Catfish, for example, smell like bloodhounds

but have lousy vision. Bass see well and don't rely on their sense of smell, while other fish rely on sound or the vibration in the water. For this reason, Jones and his fellow researchers have worked on creating specific lures for different species.

So even if earthworms wouldn't be the first choice of any discerning fish, the humble earthworm retains its appeal to fishermen. There's always room for bait that's always available, seriously cheap, and that fish will lunge at, even if it would recoil at the thought of risking its life for a humble worm.

What Exactly Are We Smelling When
We Enjoy the New-Car Smell?

D o you remember the grand old days of new-car smell that
we chronicled in *Do Penguins Have Knees?* We wrote that
the elements of that enticing aroma were paint, primer,
plastic and vinyl materials in the car, whatever material
constituted the carpeting, trim, and upholstery of the interior, and
the adhesives that held them in place.

Little did we know that only a few years later, Cadillac, a division of General Motors, would be working on a way to assure that
every car had a pleasing, uniform smell. Armed with the results of
chemistry labs and focus group research, G.M.'s Cadillac division
launched Nuance, an aroma designed to appeal to consumers on the
fence about whether to spring the extra bucks for its luxury brand.
In a 2003 *New York Times* story about how carmakers were engaging

all the senses of consumers, G.M.'s James T. Embach remarks: "You pay the extra money for leather, you don't want to smell like lighter fluid. You want it to smell like a Gucci bag." The trend is accelerating, with Porsche now introducing its own proprietary scent.

Meanwhile, the American automobile manufacturers' biggest rival has been concerned not so much with adding smells as it is with eliminating the ones already there. Recent research indicates that "volatile organic compounds," the chemicals that leach from the plastics and vinyl found in cars, may be a serious health hazard, at least in the first six months or so of the automobile's use. The big five Japanese automakers vow to reduce these emissions, even if the end result is a no-car smell.

Why Do Mosquitoes Seem To Like Some People More Than Others?

Although citing ambient temperature and visual cues as minor factors, all of the sources we cited in *What Are Hyenas Laughing At, Anyway?* agreed that mosquitoes were attracted to humans whose fragrance attracted them—sort of the insect equivalent to new-car smell. In early 2005, Rothamsted Research, in Hertfordshire, England, announced that its researchers found that bad smells can drive out good. Some folks are lucky enough to give off more than ten separate chemical compounds, "masking odors," that either repel mosquitoes or prevent the critters from detecting the human smell that they ordinarily like so much.

This research confirms previous research on cattle. By taking individual cows with masking odors away from the herd, scientists found that mosquitoes would flock to the remaining cattle in greater numbers. We've known some humans who can clear a room of other humans with their odor, but mosquitoes and humans don't seem to share the same taste in fragrances.

Why Do We Wave Polaroid Prints in the Air After They Come Out of the Camera?

n *How Does Aspirin Find a Headache?*, we gently chided Polaroid print flappers for continuing a ritual that no longer helped hasten the print's development. Since we wrote about the futility of print flapping, two important developments have occurred. The world was bombarded with Outkast's "Hey Ya," in which Andre 3000 intones the immortal lyrics: "Shake it, shake it like a Polaroid picture, shake it, shake it."

And in a stern rebuke to Outkast's exhortation, Polaroid has taken an official stance against gratuitous shaking. Its online support assures customers that a Polaroid print now dries behind a plastic window—the print itself is never exposed to the air. Indeed, the potential price for excessive flapping of prints is high, just as it can be for excessive shaking of one's booty: "Rapid movement during development can cause portions of the film to separate prematurely, or can cause 'blobs' in the picture."

Why Are U.S. Elections Held on Tuesday?

T he first sentence of our answer in *Why Don't Cats Like To Swim?* was: "Reformers are calling for weekend elections in order to increase voter turnout." Twenty years later, reformers are still calling for weekend elections in order to increase voter turnout. But national elections, and the vast majority of state and local elections, remain on Tuesday.

When we were researching this Imponderable, none of the politicians or election officials we spoke to had the slightest idea why elections were held on Tuesday (we found the answer from a historian who specialized in U.S. elections). But evidently *Imponderables* isn't required reading in political science classes. In 2000, *The Wall Street Journal* published a story by John Harwood, with the headline: "Old Election Secret Is Revealed: Why We Do It on a Tuesday—Tradition

Is Tied to Harvests And Horse Carts: Pressure Is Rising for Some Change." Although Harwood found plenty of local officials who wanted to change the system, either by allowing weekend voting (most European countries, for example, conduct elections on Sunday) or Internet voting over a period of time, Americans are resisting dumping Tuesday as voting day, even if they have no idea why. Harwood quotes Phil Kiesling, the former Oregon Secretary of State, who ushered in a vote-by-mail system that allows Oregonians to mail in ballots any day of the week during a two-week period:

> [Without an election, Tuesday] just lies there, a bit lonely.
> Tuesday is kind of a forlorn day of the week. Give Tuesday
> its due.

Why Can't We Tickle Ourselves?

Finally, scientists have set their priorities right and are studying really important stuff. Neurologists at the University College of London hooked up volunteers to a magnetic resonance imaging machine to see whether they could detect a difference between when the human guinea pigs were tickled by a machine versus when they tickled their own palms.

As we detailed in *Why Do Dogs Have Wet Noses?*, Freud argued that surprise was a crucial element in an effective tickle. Neurologists already knew that the cerebellum predicts what the effect of a particular movement will be on the rest of the body, assisting balance and locomotion. But the experiment indicated that the cerebellum is unsuccessful in warning the other parts of the brain when the stimulus is external. When the scientists controlled the timing, the machine successfully tickled the volunteers, even though the subjects anticipated it.

The subjects were also asked to activate a robot to tickle them. When they did, not only was there no tickling sensation, but several parts of the brain, including the cerebellum, acted differently from when the volunteers were surprised by the tickling. When the scientists built in a slight delay, so that the robot tickled the subjects later than the volunteers anticipated, the tickling sensation was back. So Freud's "surprise theory" is confirmed. A tickle is only a tickle if the cerebellum can't predict it. Sarah-Jane Blakemore, leader of the experiment, quipped: "So it is possible to tickle yourself, but only by using robots."

Why Aren't There Any Miniature House Cats?

When we answered this Imponderable in *Are Lobsters Ambidextrous?* almost fifteen years ago, we noted that cats are much less "plastic" genetically—it was far harder to change the size and shape of cats through selective breeding than dogs. We also quoted cat fanciers who claimed that there was little demand for miniature cats.

Times have changed. While it's sometimes hard to believe that a Saint Bernard and a Chihuahua are both from the same planet, let alone relatives, the variances between average-sized and "miniature" cats is relatively small. Perhaps the most popular of the novelty breeds is the "munchkin." A Louisiana woman, Sandra Hochenedel, found a female cat with extremely short legs living under her truck in the early 1980s. Hochenedel discovered that the cat, whom she named Blackberry, was pregnant, and in her first and subsequent litters, Blackberry passed along the short-legged trait to about half of her offspring. Munchkins seem able to run and climb adequately, but don't have the jumping ability of their long-legged peers. Another natural breed is the Singapura, known as a "drain cat" in its native land (they lived in the culverts of Singapore)—healthy female Singapuras grow to only four to six pounds. Several American breeders specialize in "downsizing" standard popular breeds, such as Persians and Siamese.

We aren't yet at the stage where Paris Hilton is carrying a "teacup cat" into nightclubs, but in another fifteen years, we're betting that a profusion of miniature cats is more likely than Paris Hilton still gracing the covers of tabloids.

Does Anyone Really Like Fruitcake?

Ever since we posed this Frustable in *Why Do Dogs Have Wet Noses?*, we've heard from readers, especially around Christmas time. A few diehards claim to love the stuff, but we will stand by the sentiments we expressed in *Do Penguins Have Knees?*: If people really liked fruitcake, wouldn't it be offered in restaurants? One perceptive reader, Fred Steinberg, compared fruitcakes to electric knives and we responded:

> Steinberg's theory is that enterprising bakers have created a food designed to be given away rather than eaten. When you think of it this way, fruitcake is the ultimate diet food, since it is never actually consumed.

One impediment has loomed over the fruitcake world, though, and has prevented fruitcakes from being marketed as a healthy treat.

In its wisdom, the U.S. Food and Drug Administration has classified fruitcake as a "heavy cake," in the same category as cheesecake and pineapple upside-down cake. The suggested serving size for heavy cakes is a hefty 125 grams, more than one-quarter pound.

The problem is that packaged fruitcake marketers are unhappy having to indicate that a single serving of their product "weighs in" at 500 calories or more. The FDA argues that its job is to determine what the "amount customarily consumed" would be. While we might argue that the customary size consumed by sane Homo sapiens is zero, the fruitcake manufacturers are lobbying for one ounce, less than one-quarter of the current size. In a letter to the FDA, Geoffrey J. Crowley, president of The Ya-Hoo Baking Co., wrote:

> Customarily, our 14-ounce loaf will serve 10 generous slices. I have never observed anyone actually eating one-third of that cake in one sitting, as called for by the present government guideline, nor would we ever suggest for anyone to do so, as we would consider that to be over-indulgent and probably unhealthy.

While we would guess that the calorie count on labels is the least of the fruitcake industry's problem expanding its market, we'll have to agree that when we think of the words "quarter-pounder," fruitcake isn't what pops into our mind.

We consider ourselves to be on Verdict Watch at Imponderables Central, and will let you know the ruling on this "weighty" subject.

Why Do Older People Tend To Snore More Than Younger People?

I n *Do Elephants Jump?*, we observed that most snoring is caused by some form of blockage of airflow during breathing. Older folks are subject to more obstructions because as we age, the muscles in the throat and mouth become flabby and protrude, blocking airflow.

All kinds of devices have been marketed to promote increased airflow in the throat. But *BMJ* (the *British Medical Journal*), in a February 4, 2006 story, announced an unlikely savior had surfaced: the Australian didgeridoo (didgeridoos, wind instruments, date back to ancient times, and were fashioned by aboriginal Australians out of tree trunks hollowed out by termites). "AS," a didgeridoo teacher, told researchers that some of his students reported decreased snoring and sleep apnea symptoms, and less daytime sleepiness, to boot. Swiss researchers decided to put twenty-five volunteers on a didgeridoorific regime, first learning proper playing techniques, and then practicing at least twenty minutes a day for a minimum of five times a week.

The results? Compared to a control group of Swiss deprived of didgeridoo playing, the folks who blew the didgeridoo did not experience increased sleep, but snored less, suffered less from sleep apnea, and felt considerably less sleepy during the day. And in good news for their relationships, the partners of the didgeridoo players reported that the snoring problem was ameliorated. Noticeably missing from the study, however, was any indication of how many partners were disturbed by the didgeridoo playing.

The researchers are not claiming that a musical instrument is a panacea for snoring problems, but the study indicates that the deterioration of muscles and tissues in the throat is not inevitable with aging.

Why Are the Sprinkles Put on Ice Cream and Doughnuts Also Called Jimmies?

Just about the same time we were writing the answer to this Imponderable in *Do Elephants Jump?*, the *Boston Globe*'s "The Word" columnist, Jan Freeman, was responding to the same question from a reader, who also wondered if "jimmy" had racist overtones. Like us, her research led her to Just Born, the candy company that many believe first made the candy. As we did, she also interviewed the current CEO, Russ Born, who reiterated that the confection was named after Jimmy Bartholomew, who manned the machine that manufactured the item for Just Born.

But Freeman also quoted literary critic and etymologist John Ciardi, who insisted in a 1986 National Public Radio commentary that the word "jimmies" was at least fifteen years older: "From the time I was able to run to the local ice cream store clutching my first nickel, which must have been around 1922, no ice cream cone was worth having unless it was liberally sprinkled with jimmies."

We heard from several readers in the Boston area arguing that jimmies were given their name in Beantown and were named after the Jimmy Fund, a charity that raises money to fight cancer, especially pediatric cancer. This theory doesn't hold water, as the Jimmy Fund started in 1948, and "jimmies" was already commonly used in the mid-Atlantic and New England areas. The confusion stems, most likely, from the fact that Brigham's, one of Boston's most popular ice cream purveyors, raised money for the Jimmy Fund by donating proceeds from jimmies sales to the charity. And where did Brigham's buy its jimmies? Just Born, of course! The Jimmy Fund derived its name not from the ice cream supplement, but from its first poster boy, a twelve-year-old cancer patient, Einar Gustafson, who was rechristened "Jimmy" for fund-raising purposes.

In *Do Elephants Jump?*, we noted that we could find no legitimate reason to think that the word "jimmy" had any racist connotation, and speculated that perhaps the belief was spawned by the confection's dark brown color. In places with both rainbow-colored and dark brown candies, the former are usually called "sprinkles" and the latter "jimmies." What we failed to note was that especially in the late twentieth century, "jimmy" became a black slang term for penis and condom. We still have found no evidence at all that "jimmy" is at all related to Jim Crow.

But we did stumble upon another piece of jimmy trivia. In the Netherlands, jimmies liberally slathered on buttered toast are a popular item. Unlike our jimmies, the Dutch use real chocolate, and you can buy milk chocolate *hagelslag* (literally, "hailstorm") or dark *hagelslag*. It may be yet another Imponderable why the Dutch think hail is hot dog–shaped while we think hail is cylindrical.

BEFORE THE LECTURE, PRUDENT NED MADE SURE TO GET BOTH HIS FLU *AND* HIS YAWN SHOT.

Why Is Yawning Contagious?

Our hero, psychologist Robert Provine, was just about the only person studying the contagiousness of yawning when we explored this Imponderable in *When Do Fish Sleep?* Provine punctured many of the myths about the causes of yawning, such as that it was caused by a lack of oxygen or an overabundance of carbon dioxide, and documented what most of us suspected—watching others yawn, listening to others yawn, thinking about yawns, and yes, reading about yawns—all lead to yawning. Yawning is as contagious as laughter or the chicken pox. We're betting you are yawning right now. If you aren't, a yawn will probably creep up on you before you finish this entry.

When it came time to theorize about *why* yawning is contagious, Provine could only speculate. His theory—that yawning might have been a way for primitive man to regulate the social interactions of groups living together, such as synchronizing sleep schedules in communal cave-dwellers—made a lot of sense, but it was impossible to prove.

In the last few years, though, there has been an explosion of research on yawning contagion, in particular from cognitive researchers. Drexel University psychologist Dr. Steven Platek and his team of researchers have tried to figure out what brain and neural pathways are involved in contagious yawning. They hooked up volunteers to an MRI to compare how subjects' brain substrates behaved when exhibiting contagious yawning versus laughing and a "neutral expressive condition." Platek and team administered several standard psychological tests, and found that those subjects who did exhibit contagious yawning showed lower levels of schizotypal symptoms and higher levels of "mental state attribution," which Platek defines as "the ability to inferentially model the mental states of others." Contagious yawners could recognize their own faces faster when their images were flashed on a computer monitor. Yawners also demonstrated more empathy for others.

Although their most recent study published in 2005 in *Cognitive Brain Research* isn't exactly beach reading ("This contrast revealed significant [FDR-corrected $P < 0.01$] activation in bilateral posterior cingulated [BA 31] and precuneus [BA 23] and bilateral thalamus and parahippocampal gyrus [BA 30] . . ."), the conclusion was clear that parts of the brain definitely acted differently when yawning spontaneously. The superior face-recognition skills of the yawners might be explained by the activation of the posterior cingulate/precuneus region of the brain, which helps process our personal memories.

Platek and his colleagues hypothesize that contagious yawning may be a primitive way for us to model our behaviors after others, albeit a totally unconscious one. Certainly, other types of animals engage in this kind of modeling behavior. When one bird of a resting flock suddenly takes wing, and the others follow, it's likely that they don't know the cause of the threat that alarmed the first bird—but imitation is a valuable survival tactic in this case. Schizophrenics and autistics, who often lack the ability to pick up social clues and score low on empathy scales, rarely yawn contagiously. Babies, who any parent can tell you, have—to put it kindly—undeveloped powers of empathy, resist yawning contagiously.

While some psychologists are poking and prodding humans, other scientists are exploring yawning behavior in other animals. Many other species of mammals, birds, fish, amphibians, and reptiles yawn. But as far as we know, the chimpanzee is the only other creature that participates in human-like contagious yawning. James Anderson, a psychologist from the University of Stirling, in Scotland, along with two Japanese colleagues, showed six female adult chimpanzees videotapes of other chimps yawning as well as other videos of chimps with open mouths who were not yawning. Two of the six subjects yawned significantly more (more than double the amount) when shown yawning videos (none yawned more when seeing videos of nonyawners). Although obviously a small sample, the one-third "success" rate is in line with—and only slightly less than—the percentage of humans that Provine and Platek found were contagious yawners. In another respect, chimps proved to be like humans: Three infants were with their mothers, and none of them yawned, even though they were watching the same videos and saw their mothers yawning.

The chimp studies excite researchers in the field because none of the other primates exhibit contagious yawning, and so far, only the chimpanzee has displayed what psychologists would describe as empathetic behavior. (Primatologist Frans de Waal's book, *Good Natured: The Origins of Right and Wrong in Humans and Other Ani-*

mals is a good place to start when exploring this topic.) Who knows? Maybe some day, this research into contagious yawning might unlock the mysteries, or at least the neurological underpinning, of empathetic and cooperative behavior.

Information about yawning is exploding on the web. Although it is a French site, Le Baillement ("The Yawn") http:baillement.com has a plethora of excellent articles in English, as well. And to view more "Gaping Maws" in the animal kingdom then you can imagine, surf thee over to http://www.gapingmaws.com/index.html.

Is There Any Logic to the Numbers Assigned to Boeing Jets? What Happened to the Boeing 717?

n *Why Do Dogs Have Wet Noses?*, we chronicled how Boeing decided to assign the 700s to its commercial transport jets, and that the marketing department decided that for its first foray, 707 had a better ring to it than 700. The 717 was skipped because the Dash 80, temporarily called the 717, lost its designation when it became an Air Force plane, known as the KC-135.

Boeing has been a beehive of 7-7 activity of late, and a stalwart *Imponderables* reader and Boeing employee, Ken Giesbers, makes sure we're updated. While the 707, 727, and 757 are currently out of production, the Boeing 777 has proved popular, particularly among international carriers; the 777-200LR Worldliner set a record for the longest nonstop flight ever (from Hong Kong, flying east, to London's Heathrow Airport).

Next on tap is the 787 Dreamliner, formerly known as the less mellifluous Boeing 7E7. In its continuing game of cat and mouse with the European consortium's Airbus, the Dreamliner is Boeing's attempt to launch the most fuel-efficient jumbo jet in the sky.

But the big news from the Imponderables front is that the Boeing 717 is back, at least for a while. After Boeing merged with McDonnell Douglas in 1996, one of the planes it acquired from MD was the 100-seat MD-95, which Boeing re-branded as the 717. While the slightly bigger 737 has been the best-selling jet in history, the 717 hasn't been competitive, and Boeing has announced that production of new 717s will end this year. You can't keep track of Boeing's sevens without a scorecard.

Why Did They Take Away Red M&M's?
Why Have They Put Them
Back Recently?

The answer to this Imponderable from *Why Do Clocks Run Clockwise?* is as dated as the clothing on *That '70s Show*. Only geezers probably remember that originally all M&M's were brown—it wasn't until 1960 that red, green, and yellow M&M's were added. The red M&M's were pulled in 1976 because of a scare about the safety of Red Dye No. 2, even though the dye was never used in the candy. In *Clocks*, we mentioned the color distribution of M&M's twenty years ago:

Color	Percent in Plain M&M's	Percent in Peanut M&M's
Brown	30	30
Yellow	20	20
Red	20	20
Orange	10	10
Green	10	20
Tan	10	0

Ten years later, we chronicled the introduction of blue M&M's in *How Do Astronauts Scratch an Itch?* and the resultant reshuffling of the color proportions in both plain and peanut M&M's.

But those days seem downright antediluvian. Since we last wrote about M&M's, almond, peanut butter, crispy, and baking bits have been introduced. And the Ph.D.'s in mathematics seem to have taken over the asylum at M&M's Brand. Forget the days of color percentages ending in zeros—that's for peasants! Here are the current color distributions for the different candies:

Color	Plain	Peanut	Almond	Peanut Butter	Crispy	Baking Bits
Brown	13	12	10	10	17	13
Yellow	14	15	20	20	17	13
Red	13	12	10	10	17	12
Blue	24	23	20	20	17	25
Orange	20	23	20	20	16	25
Green	16	15	20	20	16	12

The upstarts have taken over! Twenty years ago, brown was the most popular color in the flagship brands, plain and peanut. Now they are tied for lowest in both. And if you add up the color combinations for all six varieties, brown escapes coming in dead last by a measly one percent (red achieves this dubious distinction). Orange and green, formerly the runts of the litter, now dwarf the number of brown and red. But look at blue—it's now the most popular color, or tied for it, in all varieties. And if brown has seen better days, look what happened to tan? It has disappeared from the M&M landscape.

Tan's disappearance is all the more startling since M&M has spread its palette even more by offering unpackaged Colorworks Chocolate Candies, M&M's in twenty-one different colors with messages that can be customized on one side (the famous "m" appears on the other side). The Colorworks colors are: white, black, silver, gold, brown, red, green, orange, yellow, blue, light blue, pink, dark green, teal, aqua green, dark blue, purple, light purple, dark pink, cream, and maroon. No tan!

DAVID FELDMAN

Why Do Horses Sleep Standing Up?

I n *Why Do Clocks Run Clockwise?*, we might have overstated how universal this phenomenon is. Yes, horses have the physiological equipment to sleep standing up, and in the wild, sleeping on all fours could provide for a quick getaway in case they were threatened by a predator.

But new research indicates that horses lie down more often than we suggested. Most horse owners and researchers have observed their horses standing while sleeping, a relaxed, passive posture for them because of the ligament and tendon structure that we detailed. But when horses enter REM (rapid eye movement) deep sleep, their legs often buckle. In the middle of the night, horses usually catch their REM sleep and lie down on their sides for two to four hours at a stretch. If they cannot spread out completely to sleep, a common affliction in stables, horses often lean against a wall or any sturdy object nearby.

The *New York Times* Q&A column tackled this Imponderable several years ago and quoted Dr. Katherine A. Houpt, a physiologist at the Cornell University College of Veterinary Medicine. Although conceding that horses sleep less in the wild, she's not so sure that they stand for defensive purposes, proposing that it

> is more likely due to the fact that they eat day and night at times of year when less feed is available.
>
> In summer [when food is more abundant] they lie down a fair amount . . .

According to Houpt, when wild horses do lie down, a single horse stays on all fours as a sentry, allowing its compatriots to catch REM zzz's.

When Do Fish Sleep?

Speaking of sleep and REMs, it turns out that some mammals are joining fish in the burning-the-candle-at-both-ends game. A team led by Jerome Siegel, head of the Siegel Lab at the Center for Sleep Research at UCLA, reported in 2005 that newborn bottlenose dolphins and killer whales don't get any shuteye in their first month of life. Perhaps in exasperation, their mothers also forgo sleep during this period.

This finding astounded sleep specialists because the prevailing theory has been that REM sleep is necessary for brain development in mammals, and that hormones crucial for growth are released during sleep. While human babies sleep like, well, babies, and their mothers take every opportunity to catch some sleep, Siegel speculates that whales and dolphins might have developed this mechanism so that the babies can evade predators when they are most at risk. But wouldn't the same be true of any animal? Siegel proposes that "in the water, there's no safe place to curl up."

When older dolphins and whales do sleep, they usually float on the surface of the water or lie on the floor. But these newborns swim continuously, and don't start sleeping as much as adults until they are four or five months old.

Unlike fish, dolphins have eyelids, so they can close their eyes when they sleep. But dolphins sleep with one eye closed. Sleep researchers have never found any proof that they experience REM sleep at all, and the best evidence is that only one hemisphere of the dolphin brain is experiencing the restfulness associated with sleep.

Just as some have suggested that fish might actually be sleeping (with their eyes open), perhaps whale and dolphin babies might be enjoying some form of sleep that we haven't identified yet. And just as fish seem to go into a trance when the water is dark, maybe these marine mammals indulge in some brief periods of sleep in the midst of swimming. Unlikely, but possible.

Why Don't Crickets Get Chapped from Rubbing Their Wings Together?

We solved this Imponderable in *When Do Fish Sleep?*, but almost twenty years later, scientists have just discovered why crickets aren't going deaf. If you think cricket chirping is loud, you are right—their singing can reach 100 decibels (louder than an idling bulldozer, slightly quieter than a leaf blower at full blast), enough to cause hearing damage to humans. How can the tiny cricket's nervous system hold up to the barrage?

Humans are not deafened by their own screaming because of a phenomenon known as "corollary discharge signaling." Neurobiologists have long maintained that when your brain signals to the muscles in your mouth and throat to speak (or scream), a copy of that signal is sent to your auditory system, so it can prepare to withstand the noise.

Two zoologists from the University of Cambridge in England, James A. Poulet and Berthold Hedwig have identified the two neurons in crickets that carry corollary signals to the auditory center. They discovered that at exactly the moment at which crickets move their forewings to chirp, their auditory neurons are inhibited, so the crickets do not respond as sensitively to the noise.

In a *New York Times* article about the discovery, reporter Henry Fountain asks Poulet what implications the study might have for humans:

> In people, corollary discharges might do more than prevent sensory overload; they might help provide a sense of self. "They might help us to distinguish when I moved my arm, as opposed to when you moved my arm for me," Dr. Poulet said. The pathway is no doubt more complex than in crickets, he said, "but it's likely that there's things like that going on—it's just that nobody's seen it.

For a short but more technical explanation of Poulet and Hedwig's work, see http://www.zoo.cam.ac.uk/zoostaff/hedwig/discharge.html.

Why Does Looking Up at the Sun Cause Many People To Sneeze?

We answered this Frustable in *When Do Fish Sleep?* The consensus almost twenty years ago was that this "photic sneeze response" was likely a hereditary condition, caused by the sun (or other bright lights, such as from lamps) irritating the nerves that control sneezing. At the time, we contacted Winnipeg, Manitoba optometrist Steven Mintz, who had been a valuable source to us in the past. He couldn't help us with an answer then, but ever vigilant, he has since mailed us an article called "The Photic Sneeze Response: A Descriptive Report of a Clinic Population," published in the *Journal of the American Optometric Association*.

Much to our shock, the article cites the first report of PSR in an English book by W. S. Watson in 1875! But very little hard research has been done on PSR. The three authors of this article conducted the second-largest study ever. They received 367 completed surveys, and 122 (just under one-third) were from "self-recognized photic sneezers" (at the high range of 25 to 33 percent that we estimated in *When Do Fish Sleep?*). Most of the people did not sneeze every time when looking at the sun. Nearly two-thirds estimated that they sneezed very rarely or, more commonly, only once in a while, when exposed to sunlight. Only 15 percent reported sneezing "almost every time."

The survey of Doctors Semes, Amos, and Waterbor also confirmed another much-asked Imponderable that we mentioned in our write-up—many PSRs are serial sneezers. When asked how many successive sneezes were caused by sunlight, only 39 percent reported one sneeze. Forty percent claimed two sneezes and 28 percent answered three. Two percent of the respondents bragged that they typically sneezed nine or more times in a row! The vast majority of the serial sneezers took zero to twenty seconds in between sneezes, but a few claimed that there was usually at least a minute between sneezes.

While most PSR subjects started "sunlight sneezing" at a young age, only 13.6 percent reported sneezing before the age of ten. Ten to fourteen was the most common starting point, and more than 10 percent said that PSR didn't start for them until after the age of thirty. These findings indicate that while there still seems to be a hereditary predisposition to PSR, it may also be an acquired trait. Smokers seem to have a slightly elevated chance of acquiring PSR, but the most likely candidates of all: folks who have a deviated septum.

Inbox

We've got mail. Lots and lots of mail. Most of it comes in the form of e-mail, and most of the e-mail comes in the form of questions. But we also receive our share of comments about what we've written, most of which are complimentary. Nothing is more boring than reading a letters section full of bouquets, so we've reserved this space for readers who would prefer to give us a thorn sandwich instead of a long-stemmed rose.

We don't have space to publish all the worthy corrections and disagreements, ranging from spotting typos to arguing the aesthetic merits of high-heeled shoes. But please keep sending in your objections to anything that you find lacking—they not only lead to corrections in future printings, but keep our egos from swelling. Let's proceed to your mail, and hope we can cling to a vestige of self-respect!

There's good news and bad news about the response to our last book, Do Elephants Jump? The good news is that for the most part you agreed with our explanations and arguments. Yay! Oh, but then there's the annoying bad news. We've mentioned that we always make at least one incredibly dumb mistake in every book. Sharp-eyed readers proved, however, that at times we can exceed this quota. Heck, we couldn't even get past an aside in the introduction to the letters section of Elephants without making a mistake. Here is what we wrote:

> Some things you can count on. The swallows will return to
> San Juan Capistrano. Every summer we will be bombarded
> with crummy sequels to movies we didn't care about in the

first place. And the Red Sox will field a promising team that will wilt in the clutch.

Are we unlucky or what? The last sentence would have worked perfectly fine for eighty-six years. But of course our gratuitous joke about the Red Sox just happened to be published in November 2004. Very quickly after publication, we heard from Hal Roberts of Bellingham, Washington:

Just finished reading *Do Elephants Jump?* Another fun read. How many responses have you received from Red Sox fans regarding your statement in the introduction to the letters section? I would imagine you would get a pretty good idea of how many readers you have in the Boston area.

Readership in the Boston area seems to be just fine, Hal, but we can gain some satisfaction in printing the correction from a Seattle Mariners fan.
We got into trouble with another aside. In a chapter on "Why Do We See Stars When We Bump Our Head?" we committed perhaps our most egregious mistake ever. We wrote: "As Jerry Leiber and Mike Stoller so eloquently phrased it in their song, 'On Broadway,' 'At night the stars put on a show for free.'" Shortly after the hardbound edition was published, Jo Ann Lawlor of San Jose, California, wrote:

"At night the stars put on a show for free" is a very nice line, and you even got part of the song title right: "On." However the song it is from is not "On Broadway" (Hey, when did that kind of star ever put on a show for free, except maybe when s/he is throwing a tantrum in some restaurant or nightclub?), but "Up on the Roof." No idea whether that one was written by the same two guys or not.

Agggh. Here's a case where we made a mistake because we love the music so much, we didn't fact-check properly. Jo Ann is right, of course. "On Broadway" is the masterpiece popularized by the Drifters and written by Leiber and Stoller. "Up on the Roof" was also a big hit for the Drifters and was covered by Laura Nyro, James Taylor, and many others, but it was written by two other geniuses from the Brill Building: Carole King and Gerry Goffin. Luckily, Jo Ann was quick enough to alert us to this mistake so that we could change it in the paperback edition of Elephants.

And Lawlor nailed us on another inaccuracy, a first as far as we can remember in Imponderables *history: we didn't understand the meaning of a word. We were discussing why loons have a problem getting airborne from either land or the water. We wrote that the common loon* "cannot alight vertically from a standstill on the water." *Lawlor comments wryly:*

> Tell ya what: anything that is in the air and stops applying
> forward power will alight real vertically, real fast. In the rest
> of the paragraph you refer, correctly, to the fact that the loon
> cannot *take off* vertically as most birds do.

We cannot tell a lie. We thought that "alight" did mean "taking off." Call it temporary insanity. Call it permanent insanity. Or just call it dumb.

Regular readers of our Inbox section will be surprised that we've made it this far without any mention of boots or shoes. Worry not. Although fewer people are writing about why ranchers hang boots on fence posts, we did receive about fifteen letters on the subject. To recap, here are the theories in hand: to shield the post from rotting during rain; to discourage coyotes and other predators; to keep foul-smelling boots out of the house; to display pride; to mark where repair work on a fence is re-

quired; to amuse themselves; to signal that someone is home; to point toward a rancher's home (in case of heavy snowfall); to keep horses from impaling themselves on posts; to point toward the nearest graveyard; to shield posts from adverse reactions to the sun; to do something with single shoes lying on the road; and to offer the boots to a less fortunate cowboy. To our vast body of wisdom, we add the theory of Joanne M. Schrader of Hannibal, Missouri:

> I was told that farmers and ranchers put old cowboy boots, cans, or other stuff on fence posts to signify that it is an electrified fence. Thus it is supposed to serve as a warning not to touch the fence.

Has there been any news on the single shoe front? Have you any doubts? We've been trying to solve why you so often find exactly one shoe on the side of the road since Why Don't Clocks Run Clockwise? The time for theories is over! We need empirical research, which is why we were excited when, on a Friday evening, we received an e-mail from Anita Trout, a librarian at the University of Wyoming:

> I have just discovered your books, having read Are Lobsters Ambidextrous? last night. I was delighted to come across the question of the single shoe. The issue hasn't disappeared at all! At this very moment, in the Cooper parking lot at the University of Wyoming, there is a single, gray, man's slipper. It has been there since at least Wednesday morning.
> I have been keeping an eye on it to see what, if anything, happens. I'll report any interesting developments, if they occur.

We responded excitedly:

You are our eyes and ears! Keep an eye on it and let us know. Has it moved at all? Will it be there on Monday or the next time you come to campus? The suspense is killing us.

Then on Monday, the devastating news:

Well, this morning the slipper has disappeared. Quite frankly, I was surprised it stayed as long as it did. The grounds crew is pretty good about keeping the campus tidy. I also expected it to be gone when I came to work on Monday as we won our home football game this past Saturday and there was a fair amount of celebration going on. So, the suspense of checking each day is over, but the mystery continues.

Nothing says celebration more than picking up slippers, evidently. Foiled again in our attempts to get to the bottom of the single shoe Imponderable!

That's the problem with single-shoe research: First hand evidence is always so elusive. That's one reason why we're always interested in hearing from perpetrators. Or would-be perps, anyway. We received a long e-mail from Ashley Odell of Manchester, Connecticut. She regaled us with stories about her family summer vacations, driving to points north in an old, beat-up Econoline van, complete with three bench rows in the back and precious few windows that would open. Ashley has six older siblings and they all piled into the Econoline along with the parental units and the maternal grandmother. And as kids on long drives tend to do, they would agitate the parents, until Dad, the designated driver, told Mom to mediate by infiltrating the back section. Theoretically, this settled the problem, but it also rewarded one of the random kids by allowing them the front passenger seat:

Invariably, whichever bratty kid got put in the front would immediately roll down the window, tilt the chair back, and put their feet up. The left foot would be placed on the dash, while the right foot would be stuck just a little ways out—you guessed it—the window.

Ever since these trips in early childhood, Ashley has had her eyes peeled. She went on a long trip from Connecticut to Minnesota, and then all the way down to Fort Worth, Texas, and noted:

Contrary to what I had believed all along, my older siblings aren't the only people who stick their right feet out car windows. I saw people doing it in Ohio, Illinois, Wisconsin, Minnesota, Iowa, Kansas, Oklahoma, and Texas. Everywhere we went, I was sure to see at least one foot dangling out a window. I noticed it the most with trucks and moving vans . . .

Since that time, I've been looking for this odd behavior when I travel by road. I-93 in Vermont? Dangling right feet. I-95 through Jersey? Dangling right feet. Alligator Alley? Dangling right feet, and also not a single alligator. I noticed it on the roads leading to Gettysburg and Valley Forge just a few months ago.

Now can I say that I've actually seen a shoe fly off someone's foot and land on the side of the road? No, I can't. But I do believe that if you gathered up all of the single shoes on roads, you would find that 90 percent of them are right shoes, and that they arrived there after slipping off the feet of weary travelers who just wanted to relax a little.

Of course, that still leaves the problem of the 10 percent of shoes designed for left feet unexplained. Unless people stick their feet out of the left windows of cars as well as the

right, I guess this will remain an Imponderable, though you
can't say I didn't try.

*A cross-cultural study is needed. If your theory is correct, then left shoes
should be found on the side of the road in England. But if folks stick
their feet out of the left window in the United States, by the established
laws of Newtonian Single Shoe Theory, wouldn't most of these shoes end
up on the left side of the road? This seems like a relatively easy thesis to
prove. We still believe that more shoes are deliberately thrown out of
windows than are "innocently" lost to sleeping, out-the-window leg-
stickers.*

*Speaking of gross things in and out of cars, it's been a few books
since we've had a scary report about cockroaches in automobiles. In* How
Does Aspirin Find a Headache?, *we answered why we don't usually see
cockroaches even in crumb-filled cars, and indicated that both cold and a
lack of liquids drives the vermin to happier hunting grounds. Here's a
distasteful exception that proves the rule and confirms a stereotype or two.
Take it away, Vinnie O'Connor from Oceanside, New York:*

Cockroaches do live in some cars, namely police cars, big-
time. It seems the eggs get in the cars from the clothing or
personal items of people who are transported in the backseat.
When the eggs hatch, there is plenty of food, especially
doughnut crumbs. Cops also tend to spill a lot of drinks, due
to the nature of the business, supplying plenty of liquids to
go with the doughnuts. As far as heat goes, the cars usually
run 24-7 and have heaters. It is not uncommon for a car to be
put out of service to be fumigated because of cockroaches.

*Other animals have a more ambiguous relationship with liquids.
We discussed why cats don't like to swim in the book of the same title.*

We argued that house cats were quite agile in the water, but shunned gratuitous swimming because their fastidiousness nature led them to figure a swim just wasn't worth all the cleaning off afterward. In other words, they are lazy. Those were fighting words to David Ardnt, Jr. of Fort Hood, Texas:

I am a cat fanatic and probably the biggest cat lover on earth. I have done a lot of research on cats. In the book *Why Do Cats Sulk?*, it says: "Cats were originally desert animals. They have only been domesticated for about 5,000 years. Unlike dogs who've been domesticated for over 10,000 years.

Dogs have had a lot more time to settle down and get used to the new lifestyle. Cats kind of still have one paw in the desert, so to speak.

It seems to us that a desert animal would be more than happy to frolic in the water if given the opportunity. Our opinion is that cats sulk to impose maximum punishment on humans.

Cats won't swim and elephants won't jump. According to James Gleick of Garrison, New York, the elephant's reluctance to leap is a good thing:

It has to do with *scaling*. It's not an accident that an elephant is fat and stubby in design. You couldn't have (with available earthly biological materials) an elephant shaped like a gazelle but scaled up in size; its bones would snap. This is because an animal's mass is proportional to the *cube* of its length (or height). That is, if it's ten times longer, it's 1,000 times heavier (more or less). But the strength of a limb doesn't go up 1,000 times when you make it ten times thicker. I think that for the same reason you couldn't design an elephant that would be strong enough to jump.

He didn't ask for the plug, but if you want to know more about scaling, we'd suggest a look at Jim's book, Chaos, *which has a fascinating discussion of scaling as it pertains to humans, animals, and weather, starting on page 107.*

Speaking of scales, macadamia nuts are fattening. Which is why, perhaps, it's a good thing that they are semi-impossible to crack open in their natural state. In Do Penguins Have Knees?, *we wrote:*

> Macadamia nuts do have shells. But selling them in their shells would present a serious marketing problem. Only Superman could eat them. According to the Mauna Loa Macadamia Nut Corporation, the largest producer of macadamias in the world, "It takes 300-pounds-per-square-inch of pressure to break the shell.

A couple of Imponderables *readers had bones to pick, or shells to crack, about this. First we heard from Katie Barnes, a third-year veterinary school student at the University of Georgia:*

> There are several animals out there who are more than happy to muster up the necessary 300 psi [pounds per square inch] of jaw pressure to break open macadamia nutshells, such as the spotted hyena (estimates from 1,000 to 4,000 psi), crocodile, and several species of sharks.

Of course, a shark is unlikely to sample macadamia nuts, unless offered some inadvertently by inebriated fishermen, but Katie was ready for our argument:

One in particular that I know not only has the capability but also the desire to do so—the very rare but very beautiful Hyacinth Macaw. My godmother raises these birds and I have had the pleasure of watching them (even the young ones) casually split these shells open with their massive beaks and then proceed to gently gnaw on their owner's earlobes or hair (yikes!). So there are animals who can and frequently do shell macadamias and eat them.

We e-mailed Katie and asked her if any other animals might be able to crack open macadamias. She replied:

I don't know of any other animal other than the Hyacinth Macaw that would eat macadamias. I know of the South American agouti (a type of lagomorph), which can gnaw through Brazil nuts and their extremely hard husks, but I don't think that even the Brazil nut can match the toughness of a macadamia nut.

Macadamia nuts are from Australia originally (not from Hawaii as the Mauna Loa Company would have us believe) and Hyacinths are from South America, so this is not a natural food of theirs, but they do eat them in captivity. I don't think any native Australian animal can manage them—they don't have any native macaws down there and their other birds are pretty wussy.

We posted Katie's contributions to our blog at Imponderables.com and heard quickly from Suzanne Shriner, one of three generations of a family-run farm and bed-and-breakfast operation in Honaunau, Hawaii. The Lions' Gate Farm includes five acres of coffee and five acres of macadamia nuts. Suzanne, who manages the farm, adds to Katie's list:

> I wanted to add a couple of animals that can manage the "hardest shell in the world." Wild boars have no problem cracking them open. In fact, they are second only to rats in the amount of crop damage done to farmers here.
>
> We also have a dog that can worry the nut open. She's broken two canine teeth this way, but that doesn't stop her from laying in the field and cracking nut after nut.

And I'll bet the dog isn't afraid to swim, either.

And dogs have been known to chase a ball or two. But not orange balls—at least if the balls hang on power lines. We wrote about why you sometimes see orange balls on power lines in Why Do Dogs Have Wet Noses? *The main reason for the balls, we argued, was to alert low-flying aircraft, and to a lesser extent, birds, to the presence of power lines. Christopher Durkin of Glendale, Arizona, dissents:*

> While the orange balls do provide excellent visibility (and sometimes may be used expressly for this purpose), it is not the reason that most of them are on power lines. And even as non-ornithologists/non-electric-company-CEOs know, I doubt the power company will spend great deals of money and risk human life in high voltage installations on lines accessible only to helicopters (in some cases), just to allow migrating birds a place to rest.
>
> The real reason these were developed . . . is to reduce or eliminate the phenomenon of standing waves in power lines. You see, sometimes the wind will catch lines just right, and

they will undulate up and down (this depends on the length of the span, the tension in the wire, and the direction and speed of the wind) . . . these lines start to wave up and down. One is going up while the one above it may be coming down, and ZAAAAPPPPP! Gotham is in the dark.

This is particularly probable where lines are likely to acquire ice or freezing rain. The freezing rain hangs down uniformly along a wire, looking much like a knife edge. The wind then turns this edge horizontal, and the ensuing effect is much like that of an airplane wing.

So these orange balls are placed on the lines to quiet the waves. The placement is determined by the same variables that determine wave frequency . . . The idea is to bust up the waves into harmonics. The harmonics are both harder to start, harder to maintain, and most importantly, have much less amplitude, so the wires are less likely to cross.

The balls are orange because, hey, as long as you are putting something up there, it may as well serve several uses—increased visibility is one of those benefits . . .

Your theory makes sense, Christopher, but we haven't been able to confirm it. All the literature from "orange ball" manufacturers talks only about aircraft angle. One of the largest, Tana Wire Markers, puts the Federal Aviation Administration guidelines on its Web site at http://www.tanawiremarker.com/faa.htm. *Nary a word about harmonics or birds.*

If the FAA isn't going to talk about birds, we will. In What Are Hyenas Laughing At, Anyway?, *we gave all kinds of highfalutin reasons why parrots bob their head. Reader Carolin Duncan thought we were overintellectualizing:*

We have owned cockatoos (a Moluccan, a sulfur-crested, and a goffin) for over 20 years, and in my experience with head bobbing, our birds do it because they have learned that they will get attention if they do it. Many birds (and cockatoos are noted for this) will do almost anything to get the attention of their people, ranging from cute to very annoying behaviors.

With our birds, when they bob their heads, they are pretty much guaranteed that one of their people will either bob back and say "Yeah, yeah, yeah," or do something else that rewards the bird's head-bobbing behavior. (Okay, so we're weird). My point is that I imagine many pet birds will engage in head-bobbing behavior (or any other behavior) because they will get human attention.

Traditionally, one way of grabbing the attention of humans, whether or not they own birds, is to send them a telegram. But after 145 years, Western Union has pulled the rug out from under telegrams, suspending service as of January 27, 2006. What's the closest approximation of a telegram? These days, it's probably a free radiogram offered by ham operators as a public service. Ham operators sent countless messages from those afflicted by Hurricane Katrina. Ben Burwell, a ham radio operator from Princeton, New Jersey, noted our discussion in When Do Fish Sleep? *about why Western Union used "STOP" instead of a period to end sentences in telegrams:*

Ham operators, in modern times, send the preamble, station of origin, date, and time first. Then we say the word, "break," followed by the address and telephone number of the addressee, followed by another break, the text, break, and the signature. We don't use periods or the word "stop." We use the character "X" as a stop. For example:

Number 745, precedence R, HXG, 8, KC20WE, 2/15/06, 2003Z, BREAK

David Feldman
P.O. Box 116
Planetarium Station
New York, NY 10024-0116
(800) 555-5555, BREAK

HI DAVID X HOW'S THE WEATHER X FROM, break

BEN BURWELL

The weather's fine, Ben. Stop!

What isn't so fine, however, is when a clock runs counter-clockwise. It isn't the way nature intended, or we never would have titled a book Why Do Clocks Run Clockwise? *In* Do Penguins Have Knees?, *a reader wrote to report just such a freaky clock. Bob Carson of Indiana, Pennsylvania, was similarly victimized but decided to confront the problem head-on:*

> I experienced this oddity with a hanging clock that was fairly large with large hands. After a few occurrences I called the power company to find out what happens when there is a power outage and whether a change of polarization could occur when the power is restored. Naturally, I was greeted with a few snickers—not the sweet kind.
>
> Anyhow, a careful analysis revealed some interesting things.

If the minute hand was beyond the twelve (in a downward motion) and before the six, when the power was restored the clock would continue "clockwise." However if the minute hand was beyond the six (in an upward motion) and before the twelve, the clock would commence to run backwards. Trial and error confirmed this by placing the minute hand at various positions and interrupting the power. My conclusion was that the weight of the minute hand gave impetus to which direction the motor would start . . .

Speaking of wrong directions, we wrote in Do Elephants Jump? *that* "all sheet music is read from left to right, even in Israel." *We've got to remember that nasty tendency to use the word "all." Brian Luense of Chicago, Illinois shot us an e-mail:*

Attached you should find a photograph I took of an Arabic language hymnal that I found in regular use in St. Savior Parish Church in the old city of Jerusalem in Israel this fall. You can clearly see that in fact the music in this book is written from right to left.

I would suspect that given the choral nature of the music, it is necessary in order to have the words being sung actually readable and correspond to the needed notes. I would guess that the same would be true for any language that reads from right to left, but it is only a guess.

We sit corrected.

But we don't sit when urinating. While discussing why the sound of running water gives people the urge to urinate in How Do Astronauts Scratch an Itch?, *we quoted a gentleman who claimed that some men needed to flush the toilet before urinating because* "They need to hear the water flowing to get going." *But Carlos Robles of the Bronx, New York, points out that often the flush is a premature and futile attempt to save time:*

> I believe that many men think they are near the end of their urination and they feel that by the time the toilet finishes flushing, they too will finish urinating. So why not flush the toilet then and not waste time? (We usually multiflush when we're in a hurry.)
>
> However, most men really aren't finished when they think they are finished—there's usually more to come. If you've seen *Austin Powers: International Man of Mystery,* you'll be familiar with the scene where Powers is reanimated after being cryogenically frozen. He begins to urinate and when everybody thinks he is done, he goes at it again.
>
> It's the same for all men. They think they're going to be done but there are small reoccurrences which they have to do after the toilet flushes, leading to multiflushing.
>
> Why don't men finish urinating in one shot?

We'll leave that answer for a possible all-urination Imponderables *book, and make an utterly smooth transition to the also less than lofty subject of manhole covers. In* When Do Fish Sleep? *we debunked many of the myths surrounding why manhole covers are round, but Kelly Mueller, who is a self-described "starving college student from Texas" and civil engineering major, read a letter from a reader chastising us, and wanted to weigh in with some support:*

> I would like to submit, for your arsenal of retribution, a few thoughts from a not so fresh mind of science. Beginning

with the most difficult to prove and ending with the most obvious answer:

1. The circular shape, as the traveling distributed load (i.e., weight of the vehicle) moves from the loading end to the unloading end, allows for the least possible or likely amount of recoil in the unfixed manhole. In other words, it shouldn't spring up from the road after a vehicle drives over the manhole lid (and because this object is circular it can be approached from any direction while anticipating similar results).

2. The circular manhole provides a sounder ratio of material area to area of supporting surface (to the incased lip that the cover rests on) than other straight-edged shapes . . .

3. Cost is always a factor in design, and upkeep or maintenance is a serious expense. Imagine the number of manholes in a given downtown zone, and then imagine having to repair each of those at least once in their lifetime (and the obstruction from the crew making the repairs)—that is, if the manhole were some other shape than circular.

The steel circular support lip for the round manhole is the most worthy shape to form into concrete, or for that matter other various types of pavement, due to the shape's resilience to withstand the continuous effects of weather . . . When the surface of the road expands and contracts with temperature change, the material searches for a release of the energy created by this cycle at the weakest area of its body. The shape of a circle holds and pushes at all 360 degrees of its area to resist this escape of energy, and the steel circular lip is also involved in this same proactive cycle, but at a different rate in respect to the conflicting surrounding pavement that could push or pull on its structure.

But because both components of the manhole, the pavement and the incased lip, are round, there are minimal effects on the integrity of the functioning part of the street. Think of trying to break an egg by squeezing it from all of its surface area. The simple round [but not circular!] egg is stronger than one would anticipate! If the manhole was any other shape, say a square or a triangle, given time with temperature change, a crack would originate from at least one of its corners and continue until the crack reaches a point of release.

Maybe we can convene a gathering of manhole cover and porthole manufacturers and settle this for good. We'd have a better chance of getting them together than a bunch of cowboys. Reader Harold Meyer disputes our contention in Do Penguins Have Knees? *that the crease atop most cowboy hats is there primarily for cosmetic reasons, and contrasted the cool cowboys' attire from the creaseless hat worn by Hoss, the character on* Bonanza:

> Your answer is way out of line. Cowboys creased their hats to make them drain towards the back when it rains. A hat like Hoss wore on *Bonanza* had no crease for cosmetic reasons but if it were really worn by a cowboy out on a range there would have been a crease on it.

Speaking of drainage of liquids, we received an e-mail from Harry Gish, a longtime pharmacy owner, who read What Are Hyenas Laughing At, Anyway? *and had some comments about why so many pharmacists stand on raised platforms:*

> The original reason pharmacists' areas are raised goes back to "olden times" and is more chemistry than pharmacy (in some countries pharmacists are still called "chemists"). Back

when pharmacists primarily compounded formulas rather than "pill rolling," it was desirable to have evacuation routes for spilled liquids (and solids cleaned off by water) which often could be hazardous, such as mercury. By having the area elevated it assured easier evacuation.

One of the rarest of things nowadays is a "compounding pharmacist"—one who will make custom formulas to doctor specifications rather than just being a "dispensing pharmacist." Besides lack of knowledge of how to compound there are fears of liability problems, along with greater governmental controls in accessing the raw ingredients.

There are a number of levels of pharmacy licensing and certification as well . . . Ask any pharmacist you know what a "grandfathered pharmacist" is. Pharmacists generally weren't licensed until 1933. At that time in most states, anyone who could claim proof of having dispensed for at least three years and had a high school diploma could apply for a license.

You don't need a license to wash windows, either, and some of you had strong feelings about an Imponderable we posed in Do Elephants Jump?—*"Why Are Newspapers So Effective in Cleaning Windows?" All of the professionals we spoke to used squeegees, but preferred newspapers to paper towels. Reader Darin Furry of San Rafael, California, was the most vociferous in his defense of the humble paper towel:*

About ten years ago, *Consumer Reports* tested window cleaners and also tested methods to clean windows. They found that newspapers were average at best, and there were much more effective ways to clean glass—paper towels, rags, squeegees, etc. Personally, I found newspapers were not very absorbent, time consuming to use, and extremely messy, so I use freshly laundered rags. Despite the myth that newspapers impart a shine, my rags leave the glass shining, streak-free, and squeaky clean. As for lint? Newspapers may be less "linty" than paper towels, but cloth does a far better job.

. . . the reader should have rephrased the question as: "Where did the idea originate that newspapers are effective for cleaning windows?" I think that decades ago, there weren't paper towels and you couldn't easily launder cloth rags (and perhaps if you had any spare cloth, you'd use it to patch clothes, not make rags out of them as we do nowadays). So newspapers were a cheap and convenient alternative . . .

Just to be clear, whenever possible, we phrase the Imponderables in our books as originally posed to us, even if the premise of the question is incorrect.

Speaking of rephrasing, we wish we could take back one little word in our first book: "animal." In our chapter about how cats can see in the dark, we said: "No animal possesses the ability to give light." We meant to write "mammals"—honest! But Charlie Orme of Knoxville, Tennessee caught us on it:

It is well known that several fish do in fact emit light (bioluminescence).

Right you are, Charlie. Not just fish, either—fireflies and glowworms are bioluminescent, and the ocean is full of such creatures, from fungi to cookie-cutter sharks.

Charlie Orme assumed we had also been properly rebuked, in our discussion of why the keys on the typewriter are arranged as they are (in

Why Don't Cats Like To Swim?*), *for not mentioning a salient feature of the typewriter keyboard, which we didn't:*

> Your discussion is all well and good, except you forgot to include that Mr. Sholes also, for the benefit of his sales marketing demonstrations, put all of the letters for the word "typewriter" in the top row.

Actually, we've never been rebuked for this omission, and we didn't know this piece of information. Very cool, although it would have been ever hipper if the creator of the most popular keyboard could have figured out what to do with the "lame duck" Q on the top row.

We have been rebuked by left-handed string players ever since we discussed why we don't see lefty string players in orchestras in How Do Astronauts Scratch an Itch? *In the chapter, we cited the potential problems of bow-crashing if lefties and righties were seated next to each other, and the expense of manufacturing or retrofitting instruments specifically for lefties (although they may look symmetrical on the outside, they are asymmetrical on the inside). In* Do Elephants Jump?, *reader Michele Myhaver argued that left-handers have an advantage playing stringed instruments, since the left hand accomplishes the more complicated actions, while the right hand only needs to "saw away" on the bow. Just before the publication of* Elephants, *we heard from a "rising, left-handed seventh grader who plays the cello," Gwendolyn Cannon of Gaithersburg, Maryland. We have a strong feeling that Gwendolyn is still rising, and in the ninth grade by now. As other readers have made clear, the audience can't tell whether string players are left-handed or not, since the overwhelming majority are taught to bow with their right hand, and are given no choice about it. As Gwendolyn put it:*

The other people don't really care if you're left-handed or not. Left-handed people are very good at adapting in a right-handed world. Right now, I am eating yogurt with my right hand and writing with my left hand and I only dropped a little bit on the table, which is better than I normally do [when eating it left-handed].

I'd also like to say that I can bow faster and do the fingerings faster than anyone else in the class, and still get a good tone out.

So there.

If Gwendolyn had been born at the beginning of the early twentieth century, she might have played the cello to accompany silent movies. Of course, chances are she wouldn't have, but then how could we create a stunningly subtle segue into the next e-mail from Rich Mitchell, a film historian, editor, director, and ex-projectionist. He read our discussion of why the countdown leader on films never seems to go down to one or zero in Why Do Clocks Run Clockwise?

The countdown leader was never intended to be shown to the public. You will note that on the rare occasion you see the leader in a theater or on television, other than when intentionally included in the program material, it is by accident.

The practice of including several feet of blank film before the official picture start goes back to the earliest days of film projection. Films were then printed (and photographed) on highly inflammable nitrate film. The projection light source was a scaled-down version of the high intensity carbon arcs formerly used in searchlights, which was focused on the projector's aperture by a concave mirror and a series of condenser lenses that generated a lot of heat as well as a very bright light. Were the film to be exposed to this heat longer than the frac-

tion of a second necessary for the persistence of vision phenomena, it would burst into flame. Thus, the film had to already be running through the projector before the dowser
separating the projection head from the light source was
opened. Early exhibitors soon realized that it was easier on the
audience's eyes in the already darkened auditorium if this
start-up footage was opaque.

The practice became standardized with the introduction of
the multireel feature and the use of two projectors. Silent
films were shipped with cue sheets, which not only contained
suggestions for the music to be played with each scene, but
notations of the action or title card at the end of each reel.
Three feet from the first frame of picture was a frame marked
"START," which the projectionist used to properly frame the
image. At the sight of the described action or title on the previous reel, he was to start the second projector and on the
count of three, make his changeover. Whenever possible,
silent film reels were ended on a title card, with the same title
at the start of the next reel so that, in theory, the audience
would not notice the change in projectors.

The three-count was based on running three feet of film at
the "official" silent speed of one foot, or 16 frames, per second. Though this standard was established in 1908, it's now
known that silent films were never consistently photographed
or projected at one speed; in fact, the sound speed of 24
frames per second is a rough average of the speeds at which
silent films were being photographed and projected in 1925.

The introduction of the countdown leader was necessitated
by the introduction of sound film. If you've ever heard a projector start up in the middle of a reel with the sound turned
on, you've heard the garbled noises and changes in pitch until

the machine stabilizes. Because the sound projectors of the time needed a longer startup time, the technology council of the Academy of Motion Picture Arts and Sciences introduced what became known as the "Academy Leader" in 1930. This leader placed the "Picture Start" frame twelve feet from the first frame of a picture with the next nine feet indicated by a frame with descending numbers at every foot. The last three feet were left opaque to accommodate the start of one-reel subjects or multireel features, but since the leaders were printed up as a standard unit, they would also be attached to the other reels.

Mitchell has much more to say about this subject and we've included his thoughts on our Web site. If you want to find out how the countdown leader continued to change in television, 16mm and 35mm, and about what the various beeps before films start mean, go to http://www .imponderables.com/countdown_leader.php.

If silent movies aren't old-school enough for you, let's go back in time a little further. How about to ancient times and the origin of the word "Xmas" to signify Christmas? We received a fascinating e-mail from Ramesh Chitnis, posing an alternate theory:

In the old Roman calendar, before Pope Gregory introduced two more months in 1552, there were ten months in the year. Leaving aside the earlier six months, the names of the other four months were September, October, November, and December [in order].

These are nearly the exact words in the Sanskrit language— *Sapt, Ashta, Navam, and Dasham* . . . You will note that "ten" is written in Roman as "X." Again, the word for month in Sanskrit is "*Mas.*" Thus "Xmas" means tenth month.

Sanskrit is the Mother of all the Indian languages. As a matter of fact, Will Durant, the great American historian and philosopher, said: "India is the motherland of our race and Sanskrit is the mother of Indo-European languages."

Speaking of mothers, we made a whopper of an error in How Do Astronauts Scratch an Itch? *Reader Colin Hall writes:*

> In your question about sticking your finger in carbonated drinks driving the bubbles down, your expert says that soap increases water's surface tension. This is not the case! This little experiment will prove it.
>
> Fill a sink with water. Then, very carefully, set a paper clip on the surface. Yes, the surface. If you put it in gently enough, it'll float. You may have to try this once or twenty times. Now put a drop of soap in the water. The paper clip sinks. The soap reduced the surface tension enough to cause the paper clip to sink.

Right you are, Colin. Our source made a careless mistake and we should have caught it.

Colin also had a comment about a letter in the same book referring to an old Imponderable—the ludicrous impossibility of opening and closing the milk carton:

> I recently took a trip to Washington, D.C. During breakfast one day, I noticed how hard the milk cartons were to open. Now, I'm from Minnesota, and I had never had problems opening them there. I'm assuming that there are regional differences in how much adhesive is put on the carton. I've never had the opportunity to visit New York City, but since it was

put into one of your books, it's a problem there. My advice: if you're sick of milk cartons being hard to open, move here.

'Nuff said! We're moving to Minnesota posthaste. Colin warned us to mention these points in our next book:

I'll be watching.

That's precisely what we're afraid of! We don't have to ask for readers to point out our dumb mistakes—they're more than happy to let us know.

But we still love you to pieces. Until we meet again, let's all stay healthy and keep our dumb mistakes to a bare minimum.

Acknowledgments

This book is already dedicated to my readers, so I don't want to swell your heads *too* much. If I thank you one more time, do you promise to keep sending me your e-mails and letters full of Imponderables, praise, and even the occasional criticism? You do? Then I still love you.

But readers don't get *all* the credit for this enterprise. It has been delightful to move across the hallway from HarperCollins to Collins Reference, especially because my editor and publisher, Phil Friedman, has been so enthusiastic and supportive. He hasn't known me *quite* long enough to know how annoying I can be.

The same can't be said for my agent, Jim Trupin, who has plenty of reasons to be irritated, but has stuck with me through thick, thin, and weird. And how many artists have to face the challenge of illustrating forlorn cabooses and baffled slices of bread? Kassie Schwan has been more than up to the challenge for twelve books.

Special thanks to Elizabeth Frenchman for her invaluable research help with several Imponderables and to Mark Sinclair for his yeoman work as Webmaster at Imponderables.com.

My friends and family deserve bouquets for putting up with me while I toiled over this book, but they'll have to be content with a crummy and considerably less expensive acknowledgment. Mucho thanks to Fred, Phil, Gilda, and Michael Feldman; Michele Gallery; Larry Prussin; Jon Blees; Brian Rose; Fraya and Eli Berg; Ken Gordon; Elizabeth Frenchman; Merrill Perlman; Harvey Kleinman; Pat O'Conner; Stewart Kellerman; Michael Barson; Jeannie Behrend;

Sherry Barson; Uday Ivatury; Laura Tolkow; Mani Ivatury Tolkow; Terry Johnson; Christal Henner; Roy Welland; Judith Dahlman; Paul Dahlman; Bonnie Gellas; James Gleick; Cynthia Crossen; Chris McCann; Nancy Schwantes; Prakash Kumar; Karen Stoddard; Eileen O'Neill; Joanna Parker; Maggie Wittenburg; Ed Swanson; Ernie Capobianco; Liz Trupin; Nat Segaloff; Mark Landau; Joan Urban; Diane Burrowes; Virginia Stanley; Marjan Mohsenin; Dennis, Heide, and Devin Whelan; Ji Lu; Alvin, Marilyn, Audrey, and Margot Cooperman; Carol Williams; Dan Fuller; Tom O'Brien; Susan Thomas; Tom and Leslie Rugg; Stinky; Sara Walker Bosworth; Matt Weatherford; Amy Yarger; Alona Amsel; Jenny Kraft; Simone Cox; Steve Narow; Julia Covino; Jack Estes; Jennifer Fish Wilson; Jeni Nielsen; and Erin Podolsky.

Special thanks to my pals at Starbucks #839 for keeping me vertical; to John DiBartolo, Annette Matejik, and my step-pals for keeping me ambulatory; to Jim Leff and Chowhounds for making sure I'm well fed; to PSML and Spectropop, for keeping the musical faith; to my Popular Culture Association pals, for getting academia right; and to Bill and Saipin Chutima, Ali El Sayad, China Bushell and David and Shirley Fuentes for feeding the soul.

I had a ball talking to hundreds of experts in every conceivable field for this book. The most fun part of my job is when I find *the* expert or experts on some field I know nothing about and hearing them talk about the subject they are passionate about. Without their willingness to share their knowledge with us, *Imponderables* would not be possible. More than in most of my books, many sources came up blank when faced with some of the Imponderables in this book (amazing how few sock manufacturers have studied up on tissue paper insertion!); there isn't room to include all of these willing experts here. Instead, we'll list all of the sources who were willing to go on the record and whose expertise led directly to answers in the book:

Lucinda Ayers, Campbell's Soup.

Rick Baldwin, Federation of Historical Bottle Collectors; Brenda Bates; Paul Bates; Ruth Bavetta; Kris Becker, Northcoast Thunderbikes; Roy Benson, Running, Ltd.; Fraya Berg, *Parents*; Dr. Jerry Bergman, Society for the Study of Male Psychology and Physiology; Dr. Andy Blockmanis, Pacific Center for Sleep Disorders; Rachel Bolton, Hallmark Gold Crown Stores; Matt Bragaw, National Weather Service; Sam Brookbank; David Brown, Cornell University; Sylvia Browne; Sandy Burton, Barbecue Industry Association; Joan Buyce, Masterfoods USA.

Bob Camara, K-Jack Engineering; Maria Milagros Castro, Orchard Management; Marie Cavanagh; Suzanne Clothier, Flying Dog Press; Prof. W. Rory Coker; Louis Coletta, Tony's Ice Cream; Lisa Comegna, Lord & Taylor.

Pamela Davis, Rak Systems, Inc.; Alison Day, Santa Barbara Olive Company; Bill Deane, Baseball Hall of Fame Museum; Jerry DeBene; John DiBartolo, Polytechnic University; Bob Dietzel; Dr. Sunny Dong; Dr. Joe Doyle; Pierre duP Fourie, Baron's Palace Hotel.

Dr. Gordon Edwards; Tracy Elbert, PCBexpress; Niles Eldredge, American Museum of Natural History; Todd Lee Etzel.

Phil Feldman; Louis Ferrara, Jr., Service Station Dealers of America; Paul Fiori, Service Station Dealers of America; Martha Fischer, Cornell Lab of Ornithology; Robert Fontana, TDK Electronics Corp.; Steve Fulkerson, Saint-Gobain Containers.

Sally Garrelts, Pure Fishing; Brian Gearhart, CO_2 Advertising; Bob Geis; Sharon Gerdes, S.K. Gerdes Consulting Dairy Management, Inc.; J. Giambroni, California Dairy Research Foundation; Ken Giesbers, Boeing Company; Joe Giordano; Todd Glickman, American Meteorological Society; William D. Gordon; Richard Gualtieri, Lord & Taylor; Lori Gunter, Boeing Company.

Robert Habel, Cornell University School of Veterinary Medicine; Hazel Harber; William R. Harts; Louise Hauck; Maurice Helou, Citgo of Lyndhurst; Prof. John Hertner; Dr. Myron Hinrichs, Hasti Friends of the Elephant; Sam Ho; Fred Holabird; Holidays in Africa; Richard H. Hopper; Brian Edward Hurst; Jason Hunsaker; Cheryl Hyde, White Dolphin Inc. and Academy Fitness.

Clay Irving.

Michael Jackson, Vogel's & Foster's; Dr. Keith Jones, Pure Fishing.

Terry Kennedy, St. Peter's College; Larry Khazzam, Echo Lake Industries, Ltd.; Karen Anne Klein; Kathryn Kranhold, *The Wall Street Journal*; Lucille Kubichek, Chihuahua Club of America; Meredith Kurtzman, Otto; Stamatios Kyrkos, Le Moyne College.

Fred Lanting; Tania Lategan, Cango Ostrich Farm; Prof. John Lawrence, University of South Florida; Thomas A. Lehmann, American Institute of Baking; John Levine; Mike Lombardi, Boeing Company; Mike Lopez, Prototron Circuits; Mort Luby, *Bowlers Journal International*.

Dr. Mansoor Madani, Center for Corrective Jaw Surgery; David Maier, Oregon Graduate Institute of Science and Technology; Ralph Marburger, Combe Incorporated; Mars, Inc.; Sgt. Sean McCafferty, New York Police Department; Greg McMillan, McMillan Running; Gloria McPike Tamlyn; Dale Miller, University of Arkansas at Little Rock; Dr. Steven Mintz; Paul Morgan, K&F Electronics; Pat Moricca, Gasoline Retailers Association of America; Stephanie Moritz, The Hershey Company; Cecil Munsey; Sean Murtha, American Museum of Natural History.

Prof. Kang-Yup Na, Westminster College; Willie Ninja; Tom Nosera; Dr. David Nye, Midelfort Clinic; David Nystrom, Eastern Mapping Center, U.S. Geological Survey.

Dr. Richard O'Brien; Kirk O'Donnell, American Institute of Banking; Digger Odell.

Christina Parker, Bruster's Ice Cream; Carroll Pellegrinelli, about.com; Dr. Irene Pepperberg, MIT; Kim Piper, Bruster's Ice Cream; Carole Price, American Ostrich Association; John Pritchard, Simi Winery; Prof. Robert C. Probasco, University of Idaho; Tom Purvis; Betty Pustarfi, Strictly Olive Oil.

James Randi; Bruce Reed, Bruster's Ice Cream; Steve Reichl, American Natural History Museum; Erroll Rhodes, American Bible Society; Joseph Richardson, Diebold, Incorporated; Ronnie Robertson; Joy Robinson, Combe Incorporated.

Frank Schifano; Tim Schmitt; Steve Scorfidi; Arthur Seeds, Barbecue Industry Association; Rory Sellers; Bill Sherrard, Long Island Lighting Company; Peter Sherry; Hezy Shoshani, Elephant Interest Group; Carole Shulman, Professional Skaters Guild of America; Dean Sommer, Wisconsin Center for Dairy Research; Dan St. Louis, Hosiery Technology Center; Gregg Stengel; Jeff Stevenson, American Essentials; LCDR J. Carl Sturm; Jack Suber, American Financial Printers; Daniel Sutton; Ed Swanson; Julie Swift, Michigan Bell.

Susan Thomas; Charlie Tomlin; Josh Trupin; J. Donald Turk, Mobil Oil Corporation; Ken Tyler.

Prof. Martin Uman, University of Florida.

Gerhard Verdoorn, Birdlife South Africa; Mike Vevea; Doreen Virtue; Paul Vossen.

Tom Wagner; F. Michael Wahl, Geological Society of America; Steve Warrington, Ostrich.com; Steve Webster, Monterey Bay Aquarium; Dianna Westmoreland; Pamela Whitenack, Hershey Community Archives; Robert Wilbur, Pet Food Institute; Richard Williams, National Weather Service; Jennifer Fish Wilson; Frederick Woodruff.

JP Yousha.

Jeff Zeak, American Institute of Baking.

And to those sources who wished to remain anonymous but provided valuable research, we thank you for your contribution to the vanquishing of Imponderability.

HELP!

Pirates can keep parrots. We prefer Imponderables. We certainly prefer solving Imponderables to cleaning up after parrots.

But we can't ponder without you, so please keep sending in your Imponderables and your comments (regardless of which direction your metaphorical thumbs are pointing). If you're the first person to send in an Imponderable we use in a book, you'll receive an acknowledgment and an autographed copy as a thank-you for your contribution.

Although we accept snail mail, we strongly encourage you to e-mail us if possible. Because of the high volume of mail, we can't always provide a personal response to every letter, but we'll try—a self-addressed, stamped envelope doesn't hurt. We're much better about answering e-mail, although we fall behind sometimes when work beckons.

Come visit us online at the Imponderables Web site, where you can pose Imponderables, read our blog, consult the master index of Imponderability (including all eleven *Imponderables* books), and find out what's happening at Imponderables Central. Send your correspondence, along with your name, address, and (optional) phone number to:

feldman@imponderables.com
http://www.imponderables.com

or

Imponderables
P.O. Box 116
Planetarium Station
New York, NY 10024-0116

Index